Against the Tranquility of Axioms

Against the
Tranquility
of Axioms

Rodney Needham
ⅠⅠⅠ

UNIVERSITY OF CALIFORNIA PRESS

Berkeley / *Los Angeles* / *London*

University of California Press
Berkeley and Los Angeles, California

University of California Press, Ltd.
London, England

Library of Congress Cataloging in Publication Data
Needham, Rodney.
          Against the tranquility of axioms.

          Bibliography: p.
          Includes index.
          1. Ethnology—Addresses, essays, lectures.
2. Classification, Primitive—Addresses, essays, lectures.
3. Kinship—Addresses, essays, lectures.   4. Ethnology—
Philosophy—Addresses, essays, lectures.   I. Title.
GN325 · N448 1983   306   82-23704
ISBN 0-520-04884-9

For Tristan

Contents

LIST OF FIGURES AND TABLES   viii

INCEPTION   ix

ACKNOWLEDGMENTS   xiii

1. Advertisement   1

2. This Is a Rose . . .   19

3. Polythetic Classification   36

4. Skulls and Causality   66

5. Reversals   93

6. Alternation   121

7. Wittgenstein's Arrows   155

BIBLIOGRAPHY   167

INDEX   177

Figures and Tables

FIGURES

1. Aranda Relationship Terminology   130
2. Section-Assignment among the Aranda   131
3. Dieri Relationship Terminology   132
4. Iatmül Relationship Terminology   133
5. Asymmetric Prescriptive System with Five Patrilines and Alternation by Genealogical Level   134
6. Structural Module in Alternating Prescriptive Systems   137
7. Diagrams of Matrilateral Cross-cousin Marriage   141
8. Interconnection of Components of Alternation   145
9. Sequences in Unison   146
10. Alternation as Direct Timelike Sequence qualified by Category   147

TABLES

1. Serial Likenesses among Descent Systems   40
2. Polytypic Grouping   46
3. Polythetic Arrangement   49

Man fragt sich: "Wo soll das enden?"

Wittgenstein

Inception

The preposition in my title sets the tone of the work. It is meant to recall the designations of a number of the books of Sextus Empiricus, and to announce thereby the skeptical aims of the essays collected into this volume.

The substantive phrase is taken from Sir Thomas Browne. It alludes to the intention to disturb (the learned doctor's word) the more or less tacit assent with which certain apparently axiomatic notions are entertained. These include the concepts of fact, value, class, cause, force, reversal, alternation, correspondence, and depiction.

The investigations presented here are thus intended to continue what Waismann has described, in memorable words that I have relied upon elsewhere, as "the quiet and patient undermining of categories over the whole field of thought" (1968: 21). Their purpose is, in his words again, to loosen rigid and constricting molds of thought; to bring us to see things in a new way unobstructed by misunderstandings. In these senses they make up a set of critiques of analytical concepts; and as such they are ancillary to my more major venture of the kind, an extended critique of a psychological concept, in a monograph (1972) on belief and the language of experience. The present exercises continue to maintain, also, the intrinsic relevance of empirical considerations—in the form of collective representations as reported by classicists, orientalists, historians, and ethnographers—to what might otherwise appear to be unanchored philosophical problems. None of the exercises is claimed to be conclusive, and indeed it is characteristic of the frame of mind in which they have been composed that such an outcome is methodically abjured. If we find ourselves led to echo Wittgenstein's query, and to ask where it will all end, the only answer that can be foreseen is that there will not ever be an end. Not until we stop thinking, at any rate, and that time may just be postponed if only we can teach ourselves to think more clearly.

The choice of topics for analysis has not been systematic, though I think that in the outcome they cohere fairly well and are of comparable importance. Two of the critiques (Chapters 3 and 4) have been published before; they are reprinted not merely because I myself find them likely to prove fundamentally useful, or even because they have withstood a professional exposure, but because it seems that other students of social facts have found them actually to procure the advantages that I had in view when I wrote them.

Not all of the essays are set at the same level of difficulty. "This Is a Rose . . ." is a written version of an introductory talk that I have given, never twice with the same details, to students early in their formation as social anthropologists; successive audiences have said they liked it, so I have written it out (though this somewhat rigidifies what should be its more quicksilver character) in one of the forms that the argument can assume. "Wittgenstein's Arrows" was at first a brief jeu d'esprit of the kind with which I try to enliven my Monday morning lectures at Oxford; those students who trudged over cobbles under the drizzle to hear what I might have contrived to say next have kindly said they liked this too, so here it has been given a more sober cast and some of the substance of a conventional demonstration.

As for the style of exposition, this also varies in the demands that it makes on the reader; at one place elementary anthropological symbols are parenthetically explained, while at another the argument ultimately presupposes some acquaintance with the technicalities called for in the analysis of prescriptive systems. Nevertheless, I trust that at least the gist of each of the critiques will be readily appreciated by any serious reader with an interest in the comparative interpretation of human experience. In this connection, the present volume does not proffer the existential implications that are tactfully indicated in my previous book *Circumstantial Deliveries* (1981); but I hope all the same that these skeptical critiques will really help others to make better sense of some kinds of propositions by which we attempt to make sense of the collective forms of experience.

Finally, a comment on the question of detail in the style of analysis may be helpful, by way of preparation for the minute particularity of certain arguments. After I had first presented "Reversals" in public, as a formal lecture, I was approached by a very senior colleague. He remarked, kindly enough, that I reminded him of a man who, intending to build a house, went and hired a cabinet-maker. I later told this to an orientalist friend, and she responded that in Japan that is exactly what one does.

Against the Tranquility of Axioms is dedicated to my beloved son Tristan, who has now for twenty-three years been at the center of my life, and its first unimpugnable justification. To his existence I owe the realization of one concept that has not been undermined by any skepticism but has instead conferred all the tranquility of a veritable axiom.

R. N.

All Souls College, Oxford
Trinity Term, 1981

Acknowledgments

"Polythetic Classification" (Chapter 3) was origi-
nally published in *Man: Journal of the Royal Anthropological Institute,* n.s. 10 (1975): 349–69; "Skulls and Causality" (Chapter
4) first appeared in *Man* 11 (1976): 71–88. I am obliged to the
editor and the Royal Anthropological Institute for permission
to reprint.

"Reversals" (Chapter 5) was delivered in part as the Henry
Myers Lecture for 1980. I am indebted to the Council of the
Royal Anthropological Institute for the compliment of election
to the Lecturership, and to the School of Oriental and African
Studies, University of London, for the handsome ambience in
which the lecture was given.

Figures 1–6 (Chapter 6) are reproduced from *Elementary
Structures Reconsidered* by Francis Korn (1973). I am very
pleased to thank the publishers and Dr. Korn for their kind
permissions. I am also particularly glad to acknowledge that
my ideas on alternation (the subject of Chapter 6) had their
beginning in an ideal collaboration with Dr. Korn, at Oxford,
when she was working out her exemplary analysis of the Iatmül system.

The essays in this book have been delivered, at various
points in their composition, as lectures at the University of Oxford, most of them within All Souls College. I am immeasur-
ably obligated to the holdings of the Bodleian Library, Oxford,

and the Codrington Library of All Souls, as also to the staffs of these establishments for their expert and unflagging help.

For advice and comments on formal notation, in connection with "Reversals" and "Alternation," I am most grateful to Edward Hussey, Francis Huxley, and Tristan Needham.

1

Advertisement

I

When Tristan is asked by Isolde why he has kept himself so distant from her, on the ship that bears them to her wedding with Mark, he explains that in his country it is the custom that the escort should avoid the bride.[1] For fear of what? she wants to know. Tristan replies: "Ask the custom!"

This was a handsome answer, for it is a sound ethnographic maxim that the native never knows why he does what his tradition requires him to do; and if he thinks he does know, what he claims to know is most likely to be nothing more than part of the custom itself. If a history is to be had, so that the custom can be traced back along its dated line of transmission, this merely pushes the problem too that much further back. If the custom has changed, this presents other problems; and if the custom has not changed, the problem is yet different again and deeper (Needham 1980a: 32–33). We can only ask the custom, there-

---

1. Richard Wagner, *Tristan und Isolde,* act 1, scene 5.

fore, and the task of comparativism since the eighteenth century has been to work out in what terms to do so.

When we try to account for social forms, whether through direct involvement or with the relative detachment of a humane discipline, we naturally tend to rely on premises that we take more or less for granted. In a formalized subject these premises are properly described as axioms. The noun comes from the Greek *axiōma*, that which is thought fitting, decision, self-evident principle; from *axioûn*, hold worthy, from *áxios*, worthy (Onions 1966: 66, s.v.). An axiom is not thought to require demonstration; it is accepted without proof as the basis for the logical deduction of other statements such as theorems. The concepts that I examine in the chapters that follow are not strictly axioms by this latter criterion, for they are isolated instruments of thought rather than propositions. Nevertheless, each is commonly resorted to as though it had an axiomatic quality in that it was self-evident or could on some other ground be taken for granted.

There are however no prior or absolute standards by which to determine what shall serve as basic concepts for the analysis of social facts. Even in a stricter science, "the point of taking certain propositions as basic or primitive is simply that we cannot demonstrate everything" (Blanché 1962: 11–12). Yet in comparativism we can demonstrate more than has yet been done; and where demonstration cannot at present be achieved, we can at least analyze concepts that are taken for basic and thus better reveal their constitution and capabilities.

II

The purpose of this preliminary chapter is to advert to various themes that run through the treatment of the topics of analysis. Two general matters call for special stress.

The first is the enigmatic operation of "change of aspect," meaning by this the capacity to discriminate in an object of thought as many connotations and uses as can be discovered or contrived. There is no technique for doing this, but it can be prepared for by the appreciation of suitably striking examples;

and simply to keep the possibility in mind can lead to a more receptive attitude of the analytical imagination. The examples themselves connect up also with the second theme. This— more marked in certain of the essays, or in some parts of them, than in others—is that there are advantages of understanding to be had by being ready to suspend, on occasion, an exclusive reliance on the rigorous and sequential procedures of traditional argumentation. While not in the least depreciating the force of logic and the uses of formal argumentative constructions, I have tried to exemplify (especially in Chapter 2) certain alternative means by which we change our minds and can perhaps induce others also to change theirs. These means include aphorisms, maxims, paradigms, and metaphor.

A recent work by Marshall Morris provides some ingenious illustrations of this approach. In his monograph on saying and meaning in Puerto Rico (1981), Morris not only supplies numerous acutely observed ethnographic cases, but also tellingly brings out their peculiar quality by resort to some striking metaphors. At one place he sums up the impact of a series of rather puzzling episodes by writing: "It is as though language were merely *floating* on experience, and [were] not connected to it" (p. 37). At another place he explains how it is that in daily life there is no alternative to following procedures of inquiry that are sure to be in error; although huge amounts of time are spent in this way, "one does not so much get things done as follow along *the fault lines in communication*" (p. 54; emphases in original). Admitted, it is the characteristic features of the episodes described that give the metaphors their force; but one can easily forget the factual details, or fail to appreciate them at their proper value, and it is the metaphor that suddenly conveys the distinct quality of the experiences. Although this flash of comprehension is prepared for by a systematic and cumulative comparison of the episodes as they are recounted, what conveys the comprehension is the metaphor, and in its own intuitive fashion; for, as Borges writes, a metaphor is not the methodical likening of two things but "the momentary contact of two images" (Borges 1965: 39).

Like the spontaneous process of change of aspect, the appositeness of metaphor cannot be deliberately contrived, and neither can the revelatory capacity of paradigms and aphorisms be purposefully applied. There is no strict method in these matters, but instead a species of imaginative incitement that affects the analytical intelligence rather as works of art affect our capacities for fantasy and sympathetic participation. One reason that they cannot deliberately be brought to bear on social facts is that we cannot foresee what their outcome may be. In a formal analysis, by contrast, or in pursuing a conventional train of argument, we can have in advance a fair idea of what the possible solutions may be; for these are implicit in the character of the data we are analyzing, or they can be inferred from the premises, or else by some other prior reduction in the range of likely results we can work in the direction of a certain kind of outcome. None of this can be looked for when aphorisms and the rest are at work: this is a condition of their potential, and part of their provocative power.

To these resources I should add what I have elsewhere called "exemplary scenes" (1981: 88–89). These affecting depictions are not propositional, though they can of course be described, but they derive their impact from largely visual incidents whose power to move an individual sensibility can be neither predicted nor wholly explained. In these cases also we can prepare our minds to recognize them, and we can accord them in demonstrations something of the importance that they acquire on first impression. But we cannot draw up rules for recognizing them, let alone designing them, and we cannot apply them in analysis in the methodical way in which we resort to abstract concepts such as duality or transitivity.

While I am pressing for the acceptance of non-argumentative means of comprehension and persuasion, I am not suggesting that we should be any less skeptical in our attitude towards them than we should be when we employ such analytical concepts as are treated in the chapters that follow. To be as open to possibilities of interpreting experience as I urge in these essays does not in the least imply a relaxation of skepticism, any more than the recognition of non-argumentative means of interpretation implies a perverse commitment to the irrational or the inarticulate. To give up an undue reliance on axiomatic commitments and other false certainties in the practice of comparativism entails all the more pressingly, indeed, that we should explore every conceivable means of discovering just what we can rely on.

III

The essays that follow "This Is a Rose . . ." can be left largely to speak for themselves, and their several arguments need not be rehearsed, but there are some additional points that may be mentioned. Chapter 3 in particular has attracted attention, and further reflection, since it was first printed. Some historical notes on polythetic classification, and especially on the notion of "family resemblances," may help to set this essay in place.

Lichtenberg wrote in 1765, or shortly thereafter, that nature does not create genera and species; she creates individuals, and it is our shortsightedness that compels us to look for similarities in order to handle a great many of these at once; and then the resultant concepts become the more inexact the larger the "families" (*Geschlechter*) that we make for ourselves (Lichtenberg 1968: 13). In 1773 he added: "Many words which express whole classes, or all the rungs of a ladder, are employed as individuals or as a single rung" (p. 210). These observations were continued, within the next year or two, with the particular remark that having only one word for "color" caused confusion; and "likewise for many things we have only a single word where we ought to have more." Therefore, Lichtenberg concludes, "we should look and see, in the case of anything, whether more words should not be created which would permit distinctions" (p. 301).

These are not only pertinent marginalia on the topic of classification, from the pen of a wonderfully ingenious man, but they have a special significance as early evidences of the idea of families of similarities. It has been well established that Lichtenberg was greatly admired by Wittgenstein (cf. Needham 1972: 113, n. 4), and these two names mark points of reference on a line of thought leading to the remarkable convergence that is noticed in Chapter 3. From one direction, that is, there has been a philosophical development from Lichtenberg to Wittgenstein; and from another direction there has been a scientific development from the botanist Adanson, a contemporary of Lichtenberg, down to the experimental psychology of Vygotsky, a contemporary of Wittgenstein, and the later applications of numerical taxonomy that are reported in my survey. This convergence, nearly two centuries after the inception of the taxonomic principle in question, effected one of the most momentous transformations in western thought, and I think that its radical consequences are only beginning to exert their characteristic influence. It will be worthwhile, therefore, to emphasize the importance of polythesis with a brief introductory statement.

The conventional definition of a conceptual class is that its members must possess certain properties in common. Vygotsky has shown however that this definition is unrealistic, and Wittgenstein that it is logically unnecessary. The resultant recognition of classificatory concepts based on family resemblances has recently led to a revision of anthropological analyses of kinship and of statements about belief. The essay that follows reports the discovery that, by a singular convergence of ideas within the space of hardly more than a decade, family-resemblance predicates had already been adduced in certain natural sciences under the designation of "polythetic classification." The methodological and experimental results of this approach are set out in Chapter 3, and a variety of consequences for social anthropology are drawn from them. A main conclusion is that comparative studies carried out in the stock classificatory terms of anthropology are subverted by the fact that they do not refer to common features but to polythetic classes of social facts. This realization calls for a radical revision of the terms of comparative analysis and of the taxonomic procedures by which collective representations and social forms are grouped for study. The suggestion that I put forward is that effective comparison may nevertheless be made practicable by reliance on a purely formal terminology of analytical concepts; and the prospect I envisage is that these concepts may permit the determination of basic predicates in the study of human affairs.

My own major deployment of the polythetic principle has been carried out in an investigation of belief (1972) and, at a number of places, in the criticism of concepts of descent and affinity (e.g., 1974a); the principle has also been resorted to in a reassessment of the definition of inner states as universals (1981: chap. 3), and indeed its further applications in comparativism are unpredictable. In the crucial setting of the interpretation of ethnographic data, also, it has already produced impressive results.

Endicott, for instance, has brought about a fundamental change in the conventional way of viewing the deities of the Batek Negritos of the Malayan peninsula. Instead of accepting the deities as named personages whose attributes might vary in a bewildering fashion from one testimony to another, depending partly on the area or group concerned and partly on the presumptions of the observer, Endicott partitions Negrito ideas about their deities into imagined actions, corporeal images, and names (1979: 199). Certain images and actions (roles) are particularly appropriate to each other, but the names of the superhuman beings are partially independent of either type of attribute. Endicott argues that this form of independent variation has been a source of great confusion in attempts to explain Negrito deities; and he urges that names should be regarded as just one component, of no more defining importance than images and roles, of the composite ideas which are the deities in question (p. 200). An essential point in his method is the realization that the named deities normally consist of "combinations" of image-role sets, and that "there are more ways to combine components of a set of (more than two) elements than there are elements in the set" (p. 201).

Essentially, what Endicott does is deliberately to renounce a conventional ethnographic taxonomy of deities whereby the name of each superhuman personage is presumed to denote, monothetically, a class of attributes proper to the deity. The consequence of this conventional procedure has been to make the disparate descriptions of the deity almost impossible to reconcile, and the individuation of deities has been made hard to discern. Endicott instead reanalyzes the various and fluctuating sets of attributes as polythetic combinations.

The unsteady accounts of deities given on any occasion are the outcome of opposed and irreconcilable "dynamic tendencies," towards consolidation and differentiation, which lead to the great variation in Negrito conceptions of their deities (p. 211); and the taxonomic principle responsible—as I should put it—is that of polythetic classification. An incidental lesson of this demonstration is that the clarification achieved is not the result of a taxonomic dodge which happens to prove effective. The outcome is attained the other way round, as it were, by virtue of an intrinsic correspondence between the principle and the very constitution of the conceptual classes under investigation. To judge by the variations in what the people themselves are reported to subscribe to, it may even be that in this case polythesis is actually a mode in which the Negritos really think about their deities.

That this method of explication is far from being a modern piece of sophistry, or at least a fashionable imitation of the ideas of Wittgenstein, can be borne out by reference back to a clear statement of the case for polythesis that was published within a few years of Lichtenberg's death. This is to be found in Dugald Stewart's general observations on the subject of "the beautiful" (1810), and it is of special weight because (unlike Adanson's prior suggestions about botanical taxonomy) Stewart's analysis is directed specifically toward collective representations, the evolution of ideas, and operations of thought. I report it here with all the more assurance in that I did not know of it when I wrote "Polythetic Classification," and in the greater expectation therefore that it may incline the reader in favor of the taxonomic principle at issue.

Stewart begins by recalling the many different senses in which the epithet "beautiful" can be used, and the flawed conclusions drawn by philosophers concerning the nature of "the Beautiful." This problem arises from

> a prejudice, which has descended to modern times from the scholastic ages;—that when a word admits of a variety of significations, these different significations must all be *species* of the same *genus*; and must consequently include some essential idea common to every individual to which the generic term can be applied (p. 214).

This principle, he continues, has been "an abundant source of obscurity and mystery in the different sciences. . . ." It is merely one instance of the "idle generalities" which have arisen from the undue influence of popular epithets on the speculations of the learned (p. 215), and which are founded on "a total misconception of the nature of the circumstances, which, in the history of language, attach different meanings to the same words" (p. 216).

The prejudice, or principle, under attack can be shown to fail by means of a formal example:

I shall begin with supposing, that the letters A, B, C, D, E denote a series of objects; that A possesses some one quality in common with B; B a quality in common with C; C a quality in common with D; D a quality in common with E;—while, at the same time, no quality can be found which belongs in common to any *three* objects in the series. Is it not conceivable, that the affinity between A and B may produce a transference of the name of the first to the second; and that, in consequence of the other affinities which connect the remaining objects together, the same name may pass in succession from B to C; from C to D; and from D to E? In this manner, a common appellation will arise between A and E, although the two objects may, in their nature and properties, be so widely different from each other, that no stretch of imagination can conceive how the thoughts were led from the former to the latter (p. 217).

The idea that this illustration conveys is that of a "transitive" expression, a designation taken from Payne Knight and corresponding to what D'Alembert had distinguished as a meaning "par extension" (p. 219). What happens is that a radical idea passes, "by slight gradations," into other senses in which the word is employed. Some such "transitions" are the result of local or casual associations; but others have their origin in "the constituent principles of human nature, or in the universal circumstances of the human race" (p. 221), and they form "a most interesting object of examination to all who prosecute the study of the human mind" (p. 222).

These comments—on the part of a writer who had already alluded in another place, incidentally, to the topic of "family-likeness" (cited in Needham 1972: 113, n. 4)[2]—not only make an etymological point about the transitions of sense by which discrete meanings accrete to an original meaning of a word, but they argue more fundamentally that the denotations of a verbal concept need express no essential idea that is common to all of its applications.

When we take together the congruent observations to this

---

2. This is an appropriate place at which to extend the catalogue of allusions to this idea (cf. chap. 3, n. 1), and from a rather curious source: Crevel, namely, in *Le Clavecin de Diderot*, ascribes a certain interplay of sexual attractions, centered on Orestes, to a "game of resemblances, of family similarities" (1966: 121).

effect uttered by Adanson in France, Lichtenberg in Germany, and Dugald Stewart in Scotland, it seems a fair inference that what we now recognize as the principle of polythetic classification may have been part of the tone of thought prevailing in Europe at the end of the eighteenth century. If an investigation into the history of the idea should bear out this inference, a likely explanation would be that these different thinkers (Adanson was a botanist, Lichtenberg a physicist, Stewart a philosopher) were individually responding to what are standard, if tacit, processes in the formation of classificatory concepts. These processes are implicit in the concepts of natural languages in any tradition, not just in the speculative concerns of western civilization, and I should contend that they are thus indeed to be ascribed to what Stewart might have agreed are "constituent principles" of the human mind. Here, then, is a prime example of the kind of discovery that is to be looked for from comparativism as the practice of an empirical philosophy (cf. Needham 1981: 28), and it is under this encouraging aspect that my own survey of polythetic classification is offered in the present volume.

Granted that the historical and analytical importance of the taxonomic principle has been established, the next question is what can usefully be done to refine our grasp or application of the concept. Barden has suggested that there may be a classificatory principle which would allow both monothetic and polythetic classification; this would come into play by the analysis not of words but of acts and operations (Barden 1976). If I am not quite convinced by his argument, this is for reasons that should find their place in a monograph on taxonomy rather than in the present advertences to the importance of polythesis.

I mention Barden's interesting proposal, however, as an example of one line of possible progress. If it is accepted that there are "two radically different ways in which a class can be constituted," namely monothetic and polythetic (Needham 1980a: 62), then a conceivable way of conflating the two methods has an obvious claim to attention. Conversely, it would be of similar importance if a further concentration on polytheticism were to discriminate a variety of formal modes in which the principle might be employed. Doubtless this has already been demonstrated in effect by the method of cluster analysis in the study of natural phenomena; but if it is accepted that numerical taxonomy cannot have the same purchase on social phenomena (as I contend in Chapter 3 below), then it is very much a question whether a discrimination of distinct modes is usefully practicable in comparativism.

As a further comment, however, I think there may well be something to gain in drawing out the consequences of a phrase devised by Rhees. In discussing Wittgenstein's comparison of games (1953: secs. 66–67) as a paradigm of family resemblances, he writes of games giving various sets of rules (Rhees 1970: 157); and of these he says, in an intriguing formulation, that they serve as "centres of variation."

Finally on this score there is a memorable paradigm which for me most excellently encapsulates the taxonomic lesson. John Wisdom (1965: 138) relates the following tale. It is apparently extremely difficult to breed lions in captivity, but there was at one time in the Dublin zoo a keeper by the name of Mr. Flood who had bred many lion cubs without losing one. Asked the secret of his success, Mr. Flood replied, "Understanding lions." Asked in what consists the understanding of lions, he replied, "Every lion is different." Wisdom comments: "It is not to be thought that Mr. Flood, in seeking to understand an individual lion, did not bring to bear his great experience with other lions. Only he remained free to see each lion for itself."

IV

Theories about head-hunting (the ethnographic subject of Chapter 4) have commonly relied on the idea that the severed head is thought to contain a mystical substance or force which emanates into the environment and thus enhances fertility and well-being.

Head-hunting may not seem particularly germane to the critique of basic concepts in comparativism, but I think the conventional explanation of the practice can be shown to reveal something of much importance in the determination of our own thought.

The idea in question, which has enjoyed a practically unchallenged acceptance throughout the present century at least, has a virtually axiomatic status, but it is nevertheless incorrect as a general explanation of head-hunting. Examination of ethnographic sources finds little convincing evidence for it, and field inquiries in Borneo have not confirmed that it is the indigenous explanation there. The conclusion to my analysis is that no intermediary factor need be postulated between taking heads and acquiring prosperity. The nexus has been misunderstood by anthropological commentators, who have interpolated a fictitious entity between the cause and the effects. What is at work instead is an alternative conception of causality, and the western failure to recognize this can be accounted for by reference to the scientific tone of thought and the mechanistic idiom in which the observers and commentators were educated. Assuming this case to have been made out, the inference I draw is that in current theories about other topics also we must be thinking about exotic social facts in ways that are comparably invalid.

The argument seems to me to stand as it is (that is, as it was first presented), and it has not been publicly resisted. I wish that the Jívaro ideology had since been more fully expounded in response to the questions raised about it (chap. 4, nn. 9 and 15), but otherwise I do not think the case calls for subsequent qualification. Perhaps on a formal score, however, one observation may be helpful. In representing the form of the standard (quasi-axiomatic) anthropological explanation, I use the little formula: $a \rightarrow (x) \rightarrow (b)$. This is apt enough in certain respects, in that it depicts the interpolation of the factor x and displays this as making the causal connection between a and b; also, it corresponds to the idea that x is conveyed or transmitted. But an alternative formula may be more apt in other respects, as follows: $x (a \Rightarrow b)$. This shows the hypothetical factor x as qualifying the entire transaction linking a and b; and it also depicts the relation, rather more exactly under one aspect, as a transformation.

V

Transformations are the theme of Chapter 5 ("Reversals") as well, only in this study under a variety of forms and in a very extensive range of institutions.

The concept of reversal, or inversion, has been steadily employed in social anthropology as though it had a basic and indisputable application; it has been made the subject of correlational propositions, and it has received monographic (also monothetic) treatment as a constant in symbolic action. Yet a skeptical reconsideration demonstrates by two methods, first substantive and then formal, that reversal does not deserve its axiomatic status and that it cannot be relied upon as a basic concept of the kind it has appeared to be.

Methodologically, a particular interest attaches to the formal part of the analysis. On the one hand, the formal treatment serves a distinct purpose and contributes a new proof; but on the other hand the results have to be qualified in at least three respects, and it has to be conceded that the formal analysis does not possess the kind of certainty or decidability that its notation appears to promise.

To contrive a formula for each kind of ethnographic case distinguished presupposes that the prior determination of the "kind" has been justifiably carried out; and then there is another species of uncertainty in the description of the distinctive features of this kind of social fact. Thus there are two different dimensions of latitude in prosecuting the formal analysis: one dimension presents difficulties in the estimation of resemblances and differences among empirical phenomena; the other presents difficulties in the accommodation of a formal notation to what are taken to be distinctive features of social facts of a certain kind. In principle, there are no grounds to presume that either of these sources of uncertainty and difficulty, and hence of vagueness, can be decisively eliminated. This means that under each rubric, substantive and formal, there intervenes the speculative factor of what is usually called intuition. In this connection there is a pertinent observation by Blanché: that a purely formal construction presupposes a corresponding activity of mind, and that even a formal axiomatic system is in effect "bounded on all sides by the domain of intuition" (1962: 63).

Now one does not have to concede the existence of a distinct faculty of intuition in order to grasp the application of this remark to the far looser procedures that characterize the formal treatment of reversals. Yet, all the same, we have no good reason to complain against the qualified results, in the present case, of formal analysis. Our task is to test the grounds of the apparently axiomatic certainty of some of the analytical concepts employed in comparativism; and if an irreducible latitude affects the conclusion that a particular concept is neither basic nor sure, still the conclusion holds, even if not to the final pitch of proof. And if the demonstration reveals that certain formal constructs, for all their critical value, are themselves at the mercy of circumstantial estimations of their cogency, then this too is an instrinsic part of a truly skeptical appraisal.

Blanché, in a discussion of the roles of definition and demonstration in axiomatics, says that the two demands of psychological efficiency (i.e., in procuring persuasion) and logical rigor "pull at times in opposite directions" (1962: 15); and he goes on to ask: "If a good demonstration means simply an argument which is effective, where are we to stop?" Well, there is really no telling where it will end, except perhaps that the terminus is likely to be arbitrary. There is an apposite story, recounted by Blanché (p.16), of a nobleman's tutor who, at the end of his tether, was determined nevertheless to get his theorem accepted, and finally exclaimed in exasperation: "Sir, I give you my word of honor!"

VI

A glamorous pupil from Hawaii once gave me, under the gray chill skies of Oxford, a long necklace from her tropic islands. It is made up of small snail-shells and the hard white seeds called Job's tears (*Coix lacryma*).

An interpretation immediately proposes itself: the snails are marine creatures, and the seeds flourish in dry sunny soil, so that their conjuction may be seen as intimating a set of opposites such as sea/land, wet/dry, perhaps lowland/upland, and so on. This is a symbolic matter, and it can be checked only by recourse to Pacific ethnography. But there is another aspect to the necklace which is purely formal: the components are arranged along the cord in the order shell–seed–shell–seed–. . . . In other words, shell and seed (together probably with whatever they stand for respectively) have deliberately been made to alternate.

A fair number of other arrangements could well have been adopted instead; for instance, shell–shell–seed–shell, or shell–shell–shell–seed–seed, and so on. Also, the number of distinct kinds of components need not have been limited to only two: three or more could have been employed, and the larger the number of components the more permutations in their arrangement would have been possible. The composition of the necklace poses two formal problems: Why are there only two components? and, Why do the components alternate?

The necklace offers a pleasing example of a problematic relation that is the subject of Chapter 6. Not that alternation appears at once as intrinsically problematic. In fact, I myself have included alternation among a number of relational terms (others cited are symmetry, transitivity, and complementarity) which, as a formal vocabulary, promise to evade the difficulties and uncertainties consequent upon using the terms of common English in comparative analysis (Needham 1974a: 16). My intention on that occasion was to prepare for a true theory, in the explication of social facts, that would be phrased in "a vocabulary of analytical concepts that were appropriate to the phenomena under consideration but would not be merely derived from them." Alternation, by reason of its formal character and its frequent recurrence in institutions of very different kinds, seemed a suitable candidate for inclusion in a formal vocabulary composed of relational terms. I acknowledged in advance that this vocabulary had not yet been operationally adapted very far to the study of institutions, nor elaborated comprehensively by reliance on explicit principles of analysis, but it seemed to me nevertheless that it had shown itself to be advantageous.

It remains true that there are social systems which can be integrally defined by relations such as symmetry, asymmetry, alternation, and so on, in such a manner that the formal analysis of their constituent relations provides the key to their normative structure also (Needham 1981: 22). But this does not mean that the concepts in question have been rigorously enough determined and suitably defined; all it proves is that formal terms have certain advantages over empirical descriptions, and that even formal terms which have only lax or tacit acceptations can still be useful in the practice of comparative analysis. However, experience with an apparently formal concept such as "opposition" has taught that we may be seriously misled if we rest content with merely pragmatic utility and fail to carry out a methodical analysis of the concept in question (cf. Ogden 1932; Needham 1980a: 51–56). We can well apprehend that the justification of a concept may in effect have no end, or at any rate none other than its utility; but we cannot understand either outcome unless we first subject the concept to a skeptical scrutiny of its constitution and the grounds of its supposed efficacy.

Such are the aims of the investigation into the concept of alternation that makes up Chapter 6 below. The results are less conclusive than the above considerations make desirable, but these themselves also reduce the expectation that any analysis of the kind can be either exhaustive or final. If I cannot claim to have turned the concept of alternation about so as to present all of its aspects, I think I can promise to have exposed some of its distinctive features and thereby to have registered an autonomous form of thought and action.

VII

In the final chapter, "Wittgenstein's Arrows," we resume the exploitation, in a sustained fashion and not just incidentally, of the process of change of aspect. The object of investigation this time is the graphic device of an arrow as used in anthropological diagrams.

The exercise is intended as a pendant to "This Is a Rose . . .," and it is meant also to share the provocative character of that essay; it is to be taken less as a demonstration than as an incitement. Even if it turns out to serve no direct technical purpose in the analysis of social facts, still it should conduce to an attitude of mind appropriate to skeptical comparison. If it has a little sting in the tail, this can convey a piquant lesson about a disparity between grandiose theorizing and trivial misrepresentation. Yet at the same time the reiterated lesson of the exercise as a whole is that a representation, such as an arrowed diagram, is indeed designedly artificial and that it depends for its utility entirely on the explicit enunciation of the conventions governing its construction. This may sound obvious enough;

but then the graphic device of an arrow looks obvious—and it is anything but that.

In resorting to diagrams, it is easy to slip into treating them as the actual objects of comprehension, as though they were concrete replicas of social facts. It calls sometimes for a deliberate effort of detachment to apprehend them for what they are, namely graphic articulations of abstractions. Yet even if we manage to do this, and avoid becoming distracted by the form or components of a diagram, there may persist the hazard of taking the abstractions also for what they are not. The chief means of avoiding this danger as well is to keep any abstraction, even the most formal, under an unremitting critical regard, never ascribing to it a fixed significance but methodically viewing it under one aspect or another, according to the uses that it may subserve. This is of course characteristic of the skeptical approach that is consistently advocated in the present array of perturbations to the tranquility of axioms; but there is also a particular tactic which can provide an almost technical aid in the prosecution of that approach.

In the majority of the essays that follow, it will be seen that special attention is paid to philological considerations; for example, the development of the verbal concept of a "thing," the etymological constituents of "opposition," the parallels in other languages to the idea of "alternation," the question of a possible subliminal significance common to the words equivalent to "an arrow" in exotic linguistic traditions. Much might be written to justify the relevance of philology to the critical employment of abstractions, especially in comparative analysis; but for the present purpose I shall merely advert to some of the chief reasons.

One reason is that an automatic recourse to etymology very effectively prevents the assumption that a given word, even one that has an apparently axiomatic fixity, should have a steady and perhaps necessary significance. Another reason is that the history of a word often reveals a past model of the world, or a scheme of certain aspects of things, which quite contrasts with its present usage. Yet another is that on occasion it may be feasible to discern, through the variegation of languages and despite the historical divisions among them, a focus of significance which may be interpreted as a natural response to certain phenomena of physical nature or of human nature. At one extreme, therefore, philological considerations can provide salutary notice of the intrinsic treacherousness of words; whereas at the other extreme a philological comparison can provide evidences for the positing of "natural kinds" and "secret sympathies" underlying man's apprehension of the world. These contrasted results have a deep import, in that they pertain to ultimate predicates in the interpretation of human experience.

The grandeur of this enterprise could seem out of scale with the little exercises of "Wittgenstein's Arrows," but this would be a delusion caused by the dimensions of an inappropriate model. If we seek out the changes of aspect under which an arrow can be viewed, and if we deliberately look through the minor conformation of the diagram towards the abstractions which are articulated by it, we shall find ourselves directly implicated in ultimate questions concerning canons of representation and conceptions of social reality. In this region of conjecture there are no axioms, and the only tranquility is the transient quietude of skepticism.

Nicht Fakten, sondern gleichsam illustrierte Redewendungen.

Wittgenstein

2

This Is a Rose . . .

I

It seems to be commonly assumed that we advance our thoughts, change our minds, and form new views on things, in the main, as a result of argument.

An argument is a connected series of statements intended to establish or to subvert a position; a good argument is one in which the connections are made by logically valid inferences and in which the substantive components are factually correct. The outcome of a convincing argument is a proposition that throws light on whatever is at issue. (The verb "to argue" is derived from IE *arg-*, to be white or bright.) If such a proposition is denied, the opponent must show that the argument is logically defective or that the presumed facts are wrong. The technique of argumentation can be taught, by means of instruction in the canons of logic and the just appreciation of facts. In the practice of argument there is seldom dispute about logical criteria, though it is common enough that there is disagreement about the validity of a particular inference. In a comparative discipline, especially, the logical character is often difficult to disengage from the interpretation of what are taken to be the facts.

The first and most fundamental rule of sociological method, declared Durkheim, is to consider social facts as things. The chief distinctive sign of a thing, in his view, is that it cannot be changed by a simple decree of the will; and in general a social fact is recognizable by the power of coercion that it exerts. Durkheim's conception of a social fact was met with some resistance and misunderstanding, and in the preface to the second edition of his *Règles* he was obliged to explain himself further. "To treat facts of a certain order as things is . . . not to classify them in some category or other of reality; it is to observe a certain mental attitude toward them." This means approaching them on the principle that we know absolutely nothing of what they are, and that their characteristic properties, like the unknown causes on which they depend, cannot be discovered by even the most attentive introspection. What is recommended therefore is an attitude of deliberate detachment in the face of social phenomena, without any presumption that we understand something intrinsic about them simply because they are social. It is not that social facts are objective things, but that we should consider them objectively as though they were things.

The notion of a "fact" is not at all simple or obvious. In recent usage, the word has meant a deed (now only in legal phrases such as "after the fact," "before the fact," etc.); something that has occurred, what has happened; truth, reality XVI (Onions 1966: 341, s.v.). It comes from the Latin *facere*, to do or make, and thence is connected, significantly, with "factitious," meaning made by or resulting from art, artificial. More fundamentally, if obscurely, it is also connected with IE *dhe*, to place or put (Buck 1949: 537). A fact is not only, if at all, an independent event or state of affairs; by the very act of observation, it is something that has been made, or made up.

An entire epistemology is implicated in the meaning that is ascribed to the word "fact," according as the phenomenal universe is taken to have an autonomous conformation or else to reflect our capacities for registering perceptions. Likewise, the notion of a "thing" is also intricate and disputable; in our own linguistic tradition, this particular verbal concept (like *chose* in

French) has only relatively recently acquired the sense of an independent entity having an objective existence.

Nevertheless, the character of a fact can still be indicated by setting it in contrast with some other concept, and the standard recourse is to distinguish a fact from a value. This latter concept, with the sense of adequate equivalent or worth, is derived from the Latin *valēre*, to be strong, healthy, effective; and this in turn comes from IE **wal*, a root expressing force or power. In the course of its development, naturally, the verbal concept of value has acquired a variety of connotations different from these and more or less divergent one from another; but in common usage the notion has retained associations with the individual, the subjective, and the unsteady which have contrasted it with the objective characteristics ideally attributed to the factual.

It is very hard to give a critical justification for such a clear contrast between fact and value, yet it has been constantly resorted to in the study of human society. A noteworthy use has been the sustained contention that sociological research ought to be "value-free." The evident burden has been that a scientific or otherwise rigorous study of social conduct and collective representations should consist in the investigation of objectively ascertainable facts, and should abjure evaluative judgements. Although this is a most contestable matter, there is patently something in the proposed contrast that needs to be taken seriously. What is not at all patent, however, is what is to count as a fact, and what is to count as a value, and in what respects there are crucial differences between them.

The standard way to seek a resolution is by resort to argument, and in one regard or another this can be effective enough; particular instances can be adduced, and their logical validity and fidelity to the ethnographic evidences can be debated. But the issue is in principle not so readily decidable; it has to do with the critique of concepts, and hence as an essentially philosophical problem it may not be finally resoluble. In this event the student of social phenomena has no proven criteria by which he can decisively separate facts from values. In practice, he will need at least to bear in mind the various arguments on the point, and will thereby remain more alert to the unreliable or the tendentious in the information that he is analyzing; but although this can be effective enough sometimes, it is a piecemeal and accretive procedure that does not answer well to the inherent difficulties of the matter. It is not that argument is profitless, but that in this case it does not procure the conclusive advantages that are looked for precisely from the strictness of traditional argumentation.

II

In the comparative analysis of social facts, perhaps more strikingly than in the scientific investigation of other phenomena, the notion of "change of aspect" can have a quite revelatory use.

Thus in the interpretation of the categories of social classification in Manggarai, western Flores, a symmetric scheme of analysis shows one aspect and an asymmetric scheme another; within the terminology, a symmetric component displays one aspect and an asymmetric component another. There is no need to represent Manggarai society, in such categorical respects, as possessing absolutely any single and fixed form; instead, "it is by an imaginative combination, and then repeated recombination, of aspects that we can grasp the possibilities of order that are latent in a complex form of social classification" (Needham 1980b: 75–76). This method leads to what I have called a "systematic but relative comprehension" (p. 76), and in the explication of social forms I think it has proved to offer distinct advantages.

Now the Manggarai case calls for some technical expertness in the analysis of relationship terminologies, a familiarity with the theoretical topic of prescriptive alliance, and a command of the ethnographic literature on eastern Indonesia; but it is not essential to possess these various kinds of knowledge in order to appreciate the notion of change of aspect and to foresee its potential. A simple paradigm has been provided by Wittgenstein in his treatment of the concept of "seeing." It makes a tangled impression, he says, and indeed it is tangled. If we look at a landscape, we see all sorts of distinct and indistinct movement; one feature makes a sharp impression, another is quite hazy. What we actually see can appear "completely ragged." We see things, too, under variable aspects, and it is possible to bring about in ourselves an effortless transition from the seeing of one aspect to the seeing of another, according to the interpretation that we put upon the object or the dimension in which we see it. "There is not *one genuine* proper case of such description" (Wittgenstein, 1953: 200).

Take as an example, Wittgenstein continues, the aspects of a triangle:

"This triangle can be seen as a triangular hole, as a solid, as a geometrical drawing; as standing on its base, as hanging from its apex; as a mountain, as a wedge, as an arrow or pointer, as an overturned object which is meant to stand on the shorter side of the right angle, as a half parallelogram, and as various other things." The only limits, indeed, to the number of aspects under which this triangle can be visualized are those of one's pictorial imagination in interpreting the triangle now as one thing, now as another. If it is asked how it is possible to *see* an object according to an *interpretation*, the answer is that it is not a "queer fact," as though something were being forced into a form it did not really fit, for no forcing takes place in order to do so (p. 200).

It is possible, moreover, to consider such a paradigm and deliberately to exploit it as an example of how to forgo a reliance on the axiomatic or doctrinaire presumption that there is indeed one genuine proper description of anything. How we actually do so is a separate question that we need not consider for the present. The point is that in resorting to a paradigm we can often unclench our minds and procure a new flexibility in our apprehension of things. More particularly, in such a case we can dispense with the formal apparatus of argumentation, and we can expand our thoughts in a latitude that imposes neither logical inferences nor fixed views on what are taken to be steady facts.

In the exposition that follows, I want to consider another paradigm, though of a verbal kind this time, and to look at the contrast of fact and value through it. The paradigm may owe something to Wittgenstein. I cannot remember from whom I learned it, or on what authority it may be connected with Wittgenstein. All I recall is that it was told to me in the context of talk about Wittgenstein, and at about the time when paper-covered stencils of his "Blue Book" and "Brown Book" were in private circulation. The subject of the paradigm is a rose, and this too makes some connection with a published interest of his. In the *Philosophical Investigations*, for instance, there are two arresting sections which consider the verbal image of the color of a rose in the dark:

> 514. And if I say "A rose is red in the dark too," you positively see this red in the dark before you.

> 515. Two pictures of a rose in the dark. One is quite black; for the rose is invisible. In the other, it is painted in full detail and surrounded by black. Is one of them right and the other wrong? Don't we talk of a white rose in the dark and of a red rose in the dark? And don't we say for all that that they can't be distinguished in the dark?

Later on, also, in considering the role of certain preconceptions, Wittgenstein looks at what we should understand by the statement "A rose has no teeth" (1953: 221–22). So there may just be some substance to the supposition that the present paradigm was originally an invention of Wittgenstein's, or at least derived from something he once said.

In any case, the paradigm "This is a rose . . ." consists of a set of five statements, lacking any commentary. What Wittgenstein would have said about them is an open question; very likely he would have said nothing, but would have left the force of the paradigm to provoke such salutary reflections as it might.

III

Here is the paradigm:

1. This is a rose.
2. This rose has five petals.
3. This rose is red.
4. This rose smells sweet.
5. This rose is beautiful.

Let us take it that statement 1 is clearly the assertion of a fact, and that statement 5 is just as clearly a judgement of value. Where, then, between 1 and 5, is the line between fact and value?

Statement 2 looks less certain than 1 as a statement of fact. There may be four petals or six; one has miscounted. Or what are identified as petals may some of them be sepals; or again, a process may not yet have quite developed into a petal, so there is a question whether it should be counted. In other words, a greater degree of judgement or personal estimation enters, as compared with the assertion in statement 1. Also, "this rose" may refer to a variety which typically has five petals, whereas this individual rose happens to have only four, or as many as ten. If, as Durkheim writes, "a judgement of value expresses the relation of a thing to an ideal" (1951: 139), then in this case it is an ideal type (the official variety) that determines the character of the statement. Is it, on this reading, a statement of fact? The objection could be raised that this interpretation is not what 2 means, and that the five petals must belong to the particular bloom in question. But the statement does have both possible meanings, and the sense in which it is taken is a matter of judgement. However, the judgement does not pertain simply to a numerical fact, but the statement is judged in a context; apart from the petals of a particular bloom, we have to take into account the taxonomy of roses, the specifications to be found in a rose-grower's handbook, and other matters of the kind.

Statement 3 is still more open to dispute by individual judgements: the rose may appear red to one person but mauvish to another, and rusty to someone else. Or the present color may be that of an immature bloom; when it is full-blown it is much darker. Again, the individual petals may differ in color; some are more red than others. Moreover, no matter what the hue of these petals, it may be that typically the variety has petals that are to be described officially as red, no matter how they may appear or vary in the eye of an individual observer; fact is subordinated to ideal.

Statement 4 invokes the sense of smell, as contrasted with color in 3, number in 2, and form in 1. If these attributes are ranked on a scale of certain determination, smell comes last, so that whether or not the rose smells sweet is very much subject to the uncertainty of individual judgement. Moreover, individual powers to discriminate smells differ greatly, and so does the verbal response to smell; what is sweet to one is acrid to another, what is "delicately fragrant" to one is "kind of nice" to another.

Statement 5 rests on aesthetic values which are notoriously various and individual. Whether or not a woman or a building or a piece of music is beautiful, and the basis on which any such judgement can be made to rest, belongs to the most uncertain and undemonstrable of value judgements.

Where, then, in the light of these considerations, is the line between a statement of fact and a statement of value?

IV

We shall be in a better position to respond to this question if we start again—only this time from statement 5: "This rose is beautiful."

There are social standards of beauty, according to which statement 5 has all the objectivity of a social fact; seen under this aspect, it expresses a judgement that is general and coercive. It could be tested perhaps by putting the issue to a sufficiently large sample of individuals within a linguistic community, and if a minority did not concede that the rose was beautiful, the correctness of the statement would still stand. Moreover, the number of individuals making up the minority could still not be taken to reflect truly the individual judgements of aesthetic value, for the responses of individuals may well be overridden by a pressure to conform with social convention; roses as objects of conventional aesthetics are—by contrast with seaweed or spinach—to be reckoned as beautiful. The dictionary specifically defines a rose as a "beautiful" flower (S.O.E.D., s.v., A.1).

This happens to be a fairly steady state of affairs, but even the incessant fluctuations of fashion in clothes and adornments make the same argument; namely, that it is tradition and the influence of society which determine what is or is not to be regarded as beautiful. Underlying the factor of social pressure, moreover, there may exist also a natural impulsion to single out the rose from among botanical growths and to ascribe to it that special visual quality that we describe as beauty. It has been a favorite flower in many lands, since prehistoric times, as we are told, and it appears in the earliest art, poetry, and legend; the Romans carried it with them wherever they pushed their empire, but other civilizations prized it before they did, and it has a prominence in many other traditions. The burden of this historical testimony, taken as it stands, is therefore that, in this respect also, statement 5 reflects a judgement of such generality and certainty as to count for a fact.

Thus similarly with statement 4: a rose is by definition fragrant—the dictionary says so—and for a flower to be recognized as a rose, it should therefore smell sweet. The fragrance of the rose, as an airborne property, has a definite chemical composition, and in this respect is a physical fact. That this property is described as a fragrance, i.e., as sweet-smelling, is a linguistic fact which is almost as objective, so far as individual judgement is concerned, as is the chemical determination. In this respect also, the reported perception of the individual is likely to be powerfully constrained by social expectation, so that what is actually perceived as rather dusty or pungent is nevertheless reported as sweet.

Yet there are at least two factors which reduce the certainty of statement 4 as compared with that of statement 5. The first is that the linguistic resources for the discrimination of scents are far less ample than is the range of expressions which permit more general assessments of beauty. The second, concomitantly, is that individual powers to detect and describe scents—even in accordance with social norms—are unsure and variable. The difference can be expressed by saying that in statement 4 there is a larger value component.

Statement 3, that the rose is red, is partly a matter of conventional assessment; the rose is of such a hue that no one in a given speech-community would contest that it was red. (The color-blind might be satisfied with a spectrographic reading.) Also, the rose may be of a variety that typically is defined as red; once identified by other criteria as belonging to that variety, its proper hue must be red. But into these judgements as well there can enter a high degree of individual evaluation, as we have seen in section III above, and if we concentrate on these the statement loses a great deal of its factual character. Only individuals can perceive colors, and if individuals are asked about the color of a rose that they see, their descriptions are likely to differ considerably one from another. The differences will depend on the capacities, training, vocabulary, and much else in the constitutions of those who perceive the rose, as well as on the purposes for which they assume the question to have been asked, their appreciation of the social setting in which they find themselves, and so on. In these respects the statement appears the less objective to the degree that they are taken into account.

Statement 2, about the number of petals on the rose, is subject to all the uncertainties that we have already rehearsed (sec. III), but now with a difference of balance. When we first considered this statement we took it as an assertion of fact, and the number of petals as clearly ascertainable; there were admittedly certain variations and qualifications to be taken into consideration, but intrinsically the statement was taken to be fairly simple, sure, and objective. When we approach it from the opposite standpoint, however, the complexities and uncertainties acquire a preponderance, and the weight is thrown instead on the circumstantial contingencies among which the petals may be counted or which unpredictably affect one or another observer. Under this aspect, it is not hard to conceive numerous situations in which statement 2 would be anything but clearly factual.

As for statement 1, finally, this may not be a statement of fact at all. "This is a rose" may be an expression of wonder or surprise; the thing in question is not a rose in form, but is meant to stand for one. This could easily happen in a heraldic sketch for a coat of arms; certain points are marked by dots or blobs, and it is explained that in the fully executed version each one will be a rose. Or we can imagine a test in psychology, for pattern-recognition, in which the subject is presented with a confusion of apparently random lines; he is challenged to detect outlines of natural phenomena in them, and he traces one pattern and identifies it as a rose. Alternatively, the subject is given a number of lists of attributes and is asked to what objects they belong; he selects one list (it includes thorns, fragrance, etc.) and says that this is a rose. In these cases, what counts is interpretation.

There are of course conventional interpretations, but in symbolism these are far removed, often deliberately, from what are usually taken as statements of fact. If one is shown a flower, one may correctly identify it as a rose, as defined botanically; but to say "This is a rose" may entirely miss the point when the rose is put to a symbolic use, and pretty certainly the phrase will not then count as an informative statement of fact. A rose sent to a mistress conveys passionate love; a rose placed on a coffin intimates grievous remembrance; a rose in heraldry can mean certain things, in commercial advertising yet others. In the ancient ceremony of the Golden Rose, the Pope blesses a rose-shaped ornament on Rose Sunday, in Lent, and presents it to some individual or community as a mark of special favor. That sent by Clement V (1264–1314) to the city of Basle is preserved in the Musée de Cluny at Paris; a visitor to the museum could look into the display case and correctly enough say "This is a rose"—but since it is patently not a botanical specimen, this statement will need to be amplified, if it is to convey an alternative meaning, by reference to the history of the institution (dating back, in fact, to before 1049), its several occasions, the motives of the donors and the deserts of the recipients, the reasons for the selection of a rose as pattern for the ornament, and so on.

With each item of explanation the complex of significances becomes more intricate, and statement 1 acquires its character as an epitome of values. If a statement is taken to be factual to the extent that it does not call for interpretation, then in this regard statement 1 is not at all the simple assertion of fact that it was at first supposed to be.

V

We have now made two progressions through the five statements in our paradigm: from 1 to 5, and from 5 back to 1. The results in each course are different: in the former, statement 1 was taken as a plain factual assertion, and 5 as an estimation of value; in the latter, 5 was taken as factual, and 1 as a complex symbolic evaluation.

These contrasting results can easily be elicited, even though we have been looking for a gradient of objective certainty. We have indeed tacitly taken certainty as the definitive attribute of the "objective"; but certainty does not guarantee objectivity, any more than objectivity guarantees certainty. There is also another factor to be considered as we run along the postulated gradient, in either direction: namely, that the statements are listed and numbered in a certain order, and the order itself conduces to the idea of a gradient. (Although I did not invent the paradigm, and therefore cannot speak for its guiding idea, it looks as though in fact it was premised on the gradient of apparent certainty that I attributed to it in the first progression.)

Let us imagine that the statements were arranged differently. Should we at once think them to be in the wrong order? We might at first think so, for in whichever direction we consider the gradient to run, there is still a gradient from fact to value in the paradigm as it is at present ordered. Also, statement 3 looks like a natural bridge, as it were, or a turning-point halfway along each progression, and this may seem to confirm the presumption that there is something intrinsically right about the given order. But the very fact that the construction placed on the paradigm can be reversed argues against a true order, and there is no immediate justification of the idea that statement 3 ("This rose is red") marks the line between fact at one end of the series and value at the other. Rather, the essential point of the exercise has been the demonstration that none of the statements in the paradigm is intrinsically either a statement of fact or a judgement of value. It is not just the construction placed on the present order that can be reversed: it is the construction placed on each statement individually, according as it is interpreted by contingent criteria of fact and value.

Whether we should regard another order in the paradigm as being a wrong order depends also on why we are arranging the statements at all. The exercise has a point; it corresponds to an interest, and so does each of the statements when taken in itself. Considered most generally, the possibilities of interpretation that are implicit in the paradigm tend to bear out Collingwood's contention that the meaning of a proposition is defined by the question to which it supplies an answer (1939: chap. 5). Questions frame interests, and when we concede that the character of each statement in the paradigm is determined by the interpretation placed upon it, this implies a context which must include both the concerns of the propounder and the anticipations of the interpreter. This critical setting exposes each of the statements to an indeterminate variety of practical and metaphorical constructions, and there will surely subsist among these no order corresponding to that of the paradigm. What justifies arranging the set of statements into a series is our own interest in the prospects of instruction to be had from adopting that arrangement.

The emphasis on the interpretation of each statement about the hypothetical rose makes a further challenge to the division between the factual and the evaluative, between objective and subjective. A factual assertion should, one supposes, retain its objective character in any language; but what happens when we translate the paradigm? Consider the rendering of "This is a rose" in a society ignorant of roses; "This rose has five petals" in a language that does not provide numerals as high as five; "This rose is red" in a vocabulary that divides the spectrum differently; "This rose smells sweet," when sweetness is not recognized as a property of smells; and "This rose is beautiful" in an aesthetics that does not generalize beauty or does not ascribe such a quality to a flower. None of these considerations is ultimately decisive, for periphrase and other means of specification could in each instance convey what at first was prevented by the lack of precise lexical equivalents; but they hint all the same at many kinds of further uncertainty in the issue of fact and value when the statements are translated into languages which do not directly accommodate their terms.

The nearest that we have got to an "objective" certainty in our paradigm has been in the resort to collective representations: e.g., a dictionary definition, a professional stipulation, a socially agreed standard. This is connected with the ethnographical maxim to rely in the first place on what the people themselves say, and to such an extent that another ethnographer will render substantially the same report. (Hence the chief reproach against the intriguing narration by Carlos Castaneda of the teachings of Don Juan—that there is no means of checking any part of the ethnography against its reported source: Castaneda 1968; cf. De Mille 1978). Such collective representations, it can be concluded, are definitively social facts; that is, their generality and their power of coercion impose them on the mind of the ethnographer as they do on the people under study. But what really is the standing of "collective representations"?

The premise, in the idea of a representation, is that the

world presents itself and that we *re*-present it. Yet the world does not simply present itself: we represent it as presenting itself. Every phenomenal judgement, whether of nature or of society, is framed by expectations and reported in concomitant categories. From the moment that linguistic agreement is achieved, a categorical frame is set upon the phenomena, whether these be roses or institutionalized forms of social life. At each point of observation a linguistic representation tacitly invokes premises and criteria that are themselves social, contestable (the idioms of other societies show this), and hence uncertain. As for the qualifier that the kind of representation in question is collective, the premises are that the representation is general and that the collectivity of which it is characteristic is well bounded. But neither of these premises is acceptable without tendentious manipulation; we never know, for any collectivity, how far an idea or value may be general, and the boundaries of such a collectivity are subject to perpetual dispute.

The expression "collective representation" does not objectively convey the intrinsic characters of a form of life, but it is itself a term of art which slickly generalizes social phenomena that instead stand in need of skeptical interpretation.

VI

The considerations just briefly adduced could be added to endlessly, and indeed there is no end to the question of fact and value to which they respond.

I do not maintain therefore that the present treatment of the paradigm is the right one, or even the most effective for any particular purpose. There is no "one genuine proper case" of such an interpretation; the main force of the paradigm, rather, is to impress on the analyst of social forms the constant and unevadable necessity for interpretation and change of aspect. This necessity is not to be formulated in any proposition or precept about the contrast between fact and value. The paradigm serves its purpose not by a progressive argument and its didactic conclusion, but by means of the provocation that it offers and the new alertness of mind that this stimulus should bring about. This salutary effect can be illustrated by an assimilation to other instances of the kind.

The story has been told (I do not remember the source) that Picasso was once reproached for distorting human features out of recognition. A portrait, his critic contended, ought to look like the person portrayed. Picasso demurred, and suggested that it was not quite so straightforward a business as that; the idea of looking just like something was a bit more difficult than the interlocutor supposed. The critic thereupon produced a photograph from his wallet, showed it to Picasso, and said: "There, you see, that's my wife, and that's what she looks like." Picasso looked carefully at the little print and asked, with a hint of surprise, "Just like that?" Confidently the critic confirmed that she looked exactly like that. "Hmm," said Picasso, "isn't she rather small?"

Descartes, in search of an ultimate proposition of which he could be absolutely sure, proposed *Cogito, ergo sum*: I think, therefore I am. Once assured of this, he passed on to devise consequent proofs of the existence of God and the reality of the physical world. He thought that his initial "cogito" could not be doubted, but modern philosophers have indeed found it possible to do so, in the first place on the ground that the activity of thinking does not entail the existence of an individual thinking subject (cf. Waismann 1968: 198). In the eighteenth century, however, Lichtenberg had already strikingly anticipated the conclusion of subsequent analyses. To begin with, incidentally, he gave an ironical twist to the Cartesian proposition by permuting it to *Non cogitant, ergo non sunt*: they do not think, therefore they do not exist (Lichtenberg 1968: 708). Now this is hardly more than a characteristic quip of that ingenious man; the effect of it is done with once the amusement is felt. But at another place, Lichtenberg turned on the *cogito* a skeptical probe that goes deep.

He was reflecting on the role of chance in the making of intellectual discoveries, and on the lightweight or minor nature of the changes in our thought that we bring about by close reasoning. Then he considered the reflective subject implied in the sentence: "I understand myself." On the one hand, that is, there is a cogitating individual, and on the other there is the individual as an object of the same individual's understanding.

This reminded Lichtenberg, as he wrote in his notebook, of what he had said at some other place: "Man nicht sagen sollte: ich denke, sondern *es denkt*, so wie man sagt: *es blitzt*": one should not say "I think" but "it thinks," just as one says "it is lightning" (Lichtenberg 1971: 501).

The phrase "es denkt" really calls for some such rendering as "there is thinking going on" or "it is the case that: thinking." Also the words "es blitzt" have no close equivalent in English; one does not normally say "it is lightning," meaning that lightning is playing or striking, but that there are flashes of lightning or "there's lightning." (In another version of Lichtenberg's aphorism, alluded to by both Borges and Waismann, the phrase is more conveniently "es donnert," it is thundering.) The laconic quality of the German, in this instance, better conveys the force of the expression; and the impact of this is to lead us suddenly to see that it is the personal pronoun which inclines us to infer the existence of a cogitating subject, whereas the facts can be conveyed in a grammatical form that does not carry this implication (cf. Needham 1972: 238–39).

Picasso's instructive witticism and Lichtenberg's philosophical aphorism are different forms of utterance, but they produce the same kind of effect. Waismann describes the aphorism as a rare flash of lucidity which startles us with its piercing paradoxicality; "the meaning gradually sinks into us, a new horizon discloses itself" (1968: 197). The sense of sudden revelation that such terse remarks can produce in the mind is cognate with the intellectual discoveries about which Lichtenberg says that they all seem to depend on a kind of accident, even those which we think we have made by intense effort (1958: 457). The standard form of effort, in which we are trained up intellectually, consists in the practice of argumentation; but it was not this that was responsible for the remarks that I have just quoted, nor does their usefulness depend on any argumentative expansion or explication. They produce their effect with the suddenness and novelty of a change of aspect, and it is this character that I want particularly to bring out in the present essay.

There are plenty of indications that thought can be advanced or transformed by ideas for which the thinkers, as supposed agents, cannot claim deliberate responsibility (cf.

Needham 1972: 237–38). Waismann (1968: 197) cites the cases of Gauss and Poincaré, and the history of science contains many others. What may not so readily be conceded, though, is a main implication that is to be drawn from these examples: namely, that if ideas can come to us in this sudden way, it may be possible to convey them to others in somewhat the same way. This is what is intimated by the two instances just presented. Picasso, with his mock-innocent query, disrupts an implicit theory about representation and the criteria of similarity; Lichtenberg, with his aphoristic declaration, exposes a grammatical bias in a philosophical proposition that is claimed to be indubitable, and in two words he subverts an entire metaphysics. Well, Picasso was a genius in the arts of representation and change of aspect; also, "not everyone is a Lichtenberg, and not everyone has something to say to surprise us out of all complacency" (Waismann 1968: 198). But all the same there are two consequences that can be extracted from their examples. The first is the plainer recognition that ideas can be radically reshaped or turned without recourse to the formal apparatus of explicit argumentation. The second is the didactic prospect of changing other people's minds by a methodical recourse to aphorisms, maxims, and provocative paradigms.

These are the chief purposes behind the exposition of "This Is a Rose. . . ." The epistemology of this paradigm is of course quite ancient, and I am not contending that any of the present considerations adds substantially to the long course of debates concerning the contrast between fact and value. What I should like to see received as the potential benefit of the paradigm is precisely not the argumentative content but the unsettling impact, various and unpredictable as this is likely to be, of the perplexities that are to be evoked by a consideration of the five plausibly connected statements about a rose. Once these statements have been run through, in any order and against whatever criteria of assessment, the paradigm will have served its end.

The outcome to be looked for is that any student of social phenomena who is tempted to assume that facts and values are clearly quite different things, and that there is a clear contrast to be drawn between them, will recall the paradigm "This Is a Rose . . ." and will at least—hesitate.

It is hazardous to think that a coordination of words
. . . can have much resemblance to the universe.

J. L. Borges

3

Polythetic Classification

Convergence and Consequences

I

"A perfect intelligence would not confine itself to one order of thought, but would simultaneously regard a group of objects as classified in all the ways of which they are capable" (Jevons 1874, 2: 349).

This dizzying postulate, by virtue of its very extremity, is a commonplace in theology; the capacity is central to that "respectful chaos of unimaginable superlatives" by which men refer to God (Borges 1965: 147). Yet for all its unsettling power the notion still proffers a false security, for it implies that a finite sense can be given to the prospect of "all" the ways there are of classifying things, whereas the truth instead is that the description of reality is in principle inexhaustible: "Of any two things whatever, there is some respect in which they can be said to resemble each other and not to resemble some third thing" (Hampshire 1959: 31). This is so even on the traditional premise asserted by Jevons: "Of every class, so far as it is correctly formed, the great principle of substitution is true, and whatever we know of one object in a class we also know of the other objects, so far as identity has been detected between them" (1874, 2: 345); in other words, "a class must be defined by the invariable presence of certain common properties" (p. 412). Logically, however, there is no such necessity; and once it is admitted that the common-feature definition of a class need not be the only possible method, the hypothetical number of "ways" in which things can be classified is multiplied over and over again.

Now we know experimentally from Vygotsky that classificatory concepts are not in practice formed by children in the way traditionally supposed in formal logic (Vygotsky 1962, chap. 5); and we have been shown analytically by Wittgenstein that verbal concepts are commonly not constructed on that pattern either (Wittgenstein 1953 and 1958). Instead, classes can be composed by means of what Vygotsky calls complex thinking: specifically, in a "chain complex" the definitive attribute keeps changing from one link to the next; there is no consistency in the type of bonds, and the variable meaning is carried over from one item in a class to the next with "no central significance," no "nucleus" (1962: 64). In a remarkable parallel, Wittgenstein, writing in the same period as Vygotsky, resorted to the image of a rope (later, in the *Philosophical Investigations*, a thread) in order to convey the same constitution of a concept: "The rope consists of fibres, but it does not get its strength from any fibre that runs through it from one end to another, but from the fact that there is a vast number of fibres overlapping" (1958: 87). Among the members of such a class there is a complex network of similarities overlapping and crisscrossing; sometimes overall similarities, sometimes similarities of detail. These features Wittgenstein termed, in a since-famous phrase, "family resemblances."[1]

---

1. For a brief commentary on the history of the "family" metaphor, see Needham 1972: 113, n. 4. To the writings cited there may be added the specific parallel presented by Nietzsche (1886: 25), who referred to the similarity of Indian, Greek, and German philosophising as literally a "family resemblance"

Thus by an intriguing convergence of psychological and philosophical analyses, reported independently[2] in Russia and England respectively in 1934, the traditional common-feature definition of a class was demonstrated to be both empirically and formally defective. What Jevons had called "the great principle of substitution" no longer held as a universal principle of classification, and a class could therefore no longer be defined necessarily by the invariable presence of certain common attributes. Since the members of a class composed by sporadic resemblances[3] were not assumed to be identical in any respect, it was no longer true that what was known of one member of a class was thereby known of the other members. What Jevons had described as a "correctly formed" class was correct only by convention; and in numerous circumstances, as both Vygotsky and Wittgenstein proved, the convention was factually incorrect.

(*Familien-Ähnlichkeit*). We know moreover that Nietzsche, like Wittgenstein, was an admirer of Lichtenberg, a notable predecessor in this train of thought; he judged Lichtenberg's aphorisms to be worth a place in a "treasury of German prose," and found them worth reading again and again. It should be remembered, all the same, that *Familienähnlichkeit* was in literary use over a century before Wittgenstein wrote the *Philosophical Investigations*. Grimm, *Deutsches Wörterbuch* (vol. 3, 1862) glosses the word with the Lat. *gentilis similitudo* (p. 1306, s.v.). Thomas Braun has directed me to the statement in the preface (vol. 1, 1854: cols. xxxix-xli), where it is explained that Latin equivalents are supplied as the best means to clarity, and has kindly given me his opinion that the phrase is not a literary tag such as would indicate a classical origin of the notion. For a good philosophical analysis of this idea, see Campbell (1965).

2. In an earlier reference to this convergence (Needham 1972: 111–14), I thought I was reporting a discovery, but I afterwards found that I had been anticipated by Stephen Toulmin (1969). Toulmin reports that he read Vygotsky in alternation with Wittgenstein's *Zettel* (1967a) and found his head ringing with intellectual echoes: "The theoretical parallels, the similarities in general attitude, even the tones of voice of the two men were too close to be entirely independent" (p. 71–72). He suggests moreover an indirect connection between Vygotsky and Wittgenstein through the psychologist Karl Bühler, who is frequently cited by Vygotsky in *Thought and Language*, was an acquaintance of Wittgenstein's sister (at whose house Wittgenstein could have met him), and was also "a major contributor to modern linguistic theory" (p. 72).

3. On the ambiguity of the idiom of "family" resemblances, see Needham 1972: 112–13. I have previously alluded to such resemblances as "serial likenesses" (1972: 119, n. 6; 1974a: 49), but on the whole the qualifier "sporadic" seems the most apt.

II

This conceptual revolution, after the millennial hegemony of the formal logic of the schools, might well have been expected to have consequences for a wider range of empirical disciplines. How far it actually did so, in one scientific undertaking or another, is a matter yet to be determined by a historian of ideas. It certainly had implications for social anthropology, however, and in 1970 an attempt was made to draw some of these out so far as they affected the analysis of kinship and marriage (Needham 1971b; reproduced in Needham 1974a, chap. 1). A point worth making about that revisionary exercise is that the approach was not deductive, but proceeded instead by way of explanatory difficulties that were commonly encountered in the empirical practice of anthropological comparison.

Among the topics dealt with, a paradigm case was presented by the concept of "descent," a notion which in both descriptive and comparative studies had led, I contended, to typological confusion. I proposed therefore the disintegration of this speciously univocal notion, in favor of formal criteria that represented logical possibilities. The result was the discrimination of six elementary modes in which rights could be transmitted from one generation to the next. An important aspect of this procedure, in connection with the theme of the present essay, is that an insufficiently discriminative taxonomic concept was replaced by a set of criteria which might be matched only sporadically, and in highly various combinations, by the jural institutions of real societies.

A consequence of conceiving descent systems in this way, I then argued, was that among a number of societies compared in any formal respect there would not be presumed to exist any empirical feature common to all: "in other words . . . they may not compose a class in the conventional sense" (1974a: 49). This was illustrated by a simple comparison (as in the present table 1) of three hypothetical societies (A, B, C), each constituted by three features ($p, \ldots, v$).[4] Let features r and t be each a type of

---

4. That three features were assigned to each society means that formally speaking the demonstration was not conducted in the simplest possible terms;

right transmitted in a given formal mode, e.g., mode 1 ($m \rightarrow m$), male to male (cf. Needham 1974a: 47). There is then a resemblance, r, between societies A and B, and another, t, between B and C, but none between A and C. Yet in ordinary anthropological practice these three societies could all be classed together as "patrilineal." "A crucial misdirection can thus be given to our thought by the uncritical employment of the received idea as to what a class is" (p. 50).

TABLE 1.   *Serial Likenesses among Descent Systems*

A	p,q,r	
B		r,s,t
C		t,u,v

I have recapitulated this example in some detail in order to provide for a comparison with others that are to follow below, and it is on the paradigm case of descent systems that I wish to focus particular attention. But I should not allow it to be thought that it is only in the analysis of kinship and marriage that there are serious disadvantages for social anthropology in subscribing to a conventional but unrealistic idea of how a class is formed. On this score, let me just allude briefly to a far larger and quite fundamental issue, namely our conception of human nature.

In the development of anthropology, the essential capacities of man, the intellectual and psychic resources that make up a common nature in mankind, have largely been taken for granted. In this century, it is true, there has been much prominent debate concerning the question whether all men reasoned alike, but at least it was never denied that reasoning was a universal human capacity. For the rest, the tacit presumption was that such capacities were well known, in advance of specific ethnographic comparison, and also that they were already adequately discriminated by European languages. Prominent among these components of human nature was the capacity for belief, and it has been a standard and unquestioned feature

---

but the logically redundant features emphasized the point, I thought, and conduced perhaps to an easier assimilation of the principle to empirical instances.

of ethnographic works that they ascribed to their subjects, i.e., to peoples belonging to quite disparate cultural and linguistic traditions, the common mental capacity of believing. It occurred to me, however, that this was in fact highly problematical, especially since philosophers in even our own tradition appeared increasingly baffled in their attempts to determine just what belief consisted in. I therefore undertook an investigation of this problem (Needham 1972) from a comparative standpoint, and came to the conclusion that (to be very cursory) the supposed capacity for belief consisted of no more than the custom, in a particular linguistic tradition, of making statements about belief.

In the course of the argument that led to this conclusion, a decisive step was the analysis of the verbal concept of belief after the fashion demonstrated by Wittgenstein in his examination of capacities and states of mind such as seeing, comparing, being guided, feeling, expecting, and so on (cf. Needham 1972: 116–19). In these investigations Wittgenstein showed again and again that a verbal concept which was taken to denote a distinct capacity or inner state actually referred instead to a range of phenomena (experiences, utterances, actions) that were linked only sporadically into a class by their family resemblances. In each case there was no common feature among the phenomenal instances making up the class; and in any particular case, even if a common feature could be discerned, there would remain the question why that feature should be the capacity in question (cf. Needham 1972: 121).

The outcome, then, was that a distinct capacity for belief had been attributed to human nature on the basis of an incorrect inference from the uses of a verbal concept; and that the crucial and underlying mistake was the uncritical acceptance of a traditional definition (e.g., that of Jevons: "the invariable presence of certain common properties") of the composition of a conceptual class. It is the recognition instead of a class composed by sporadic resemblances that makes the methodological connection between the analysis of descent and the analysis of belief, between the comparative study of jural institutions and of human capacities. The present essay is intended to establish yet more firmly the taxonomic principle in question, and to indicate certain general and inescapable consequences of applying the principle in social anthropology.

III

___

When I wrote my "Remarks," I was inspired directly by the writings of Wittgenstein, and I thought I was making an original application of his ideas to the practice of an empirical discipline. Certainly the venture was an independent one, and original within social anthropology, but it proved not to have been the first exercise of the kind. The encouragement to be had from this fact may be underlined if I state the circumstances of this realization.

At the University of California, where I read my critique of kinship and marriage to a graduate class in 1971, it was mentioned to me that in zoology also a taxonomy existed in which classes of creatures were grouped by what were in fact family resemblances. Such classes, I was told, were termed "polythetic."[5] So there were natural scientists, it appeared, who in their own field had already broken away from the scholastic tradition in classification of the phenomena they studied. This was very interesting, and a subsequent investigation thoroughly confirmed an instructive convergence of taxonomic method. I present here the main points established, firstly in the hope that the prior recognition of polythetic classes, on the part of different sciences, may induce social anthropologists to take more seriously the relevance of the taxonomic principle to the study of social facts. In the second place, the convergence seems to me to deserve notice for the sake of its intrinsic significance in the history of ideas.[6]

___

5. I am very grateful to Mrs. Aileen Garsson Baron, then a graduate member of the Department of Anthropology at the Riverside campus of the University of California, for being thus "the onlie begetter" of this ensuing study.

When I returned to Oxford I remembered the term and asked occasionally about it, of biologists and others, but drew blanks. The matter then receded from my attention until December 1974, when Dr. R. N. Pau, of the Department of Zoology at Oxford, delivered a seminar paper to social anthropologists on the subject of primitivity, and hence taxonomy, in zoology. When I asked him about polythetic classes he very kindly supplied me with two important references (Sokal and Sneath 1963; Mayr 1969) which laid the basis for the present article.

6. On the historical count, I think it sufficient for the present purpose to make only a summary survey of certain of the sources. There remain a great

The starting point is the work of the great French botanist Michel Adanson, who proposed that a member of a class of plants did not need to possess all the defining features of the class, and that a deviant specimen did not need to be assigned to a separate class (Adanson 1763, 1: cliv sqq.).[7] "The important point he made was that creatures should be grouped together on the greatest number of features in common, and there is no justification for deciding *a priori* on the relative importance of characters in making a natural taxonomy" (Sneath 1962: 292).

In classical taxonomy, nevertheless, taxa were in fact usually described by specifying features that were absolutely diagnostic for the class of things considered. "Linnaeus constantly revised his differentiae when the characters of newly discovered species showed that a character of a previously known taxon was no longer exclusive to that taxon" (Mayr 1969: 82), and this was a recurrent difficulty encountered generally by adherents to the common-feature principle of classification. But after 1859, and under the influence of the theory of evolution by natural selection, the definition of a class changed: "the definition of the logicians—'individuals sharing common characters'— was replaced by 'members of a group having descended from a common ancestor'" (Mayr 1969: 83). It was then no longer necessary to stipulate that the members of a taxon should invariably possess the definitive features in common; for "there is no assurance that any given character of an ancestor will persist in all its descendants" (Sokal and Sneath 1963: 217).

In the eighteenth and nineteenth centuries, therefore, the grounds for a revision of taxonomic premises were already well laid; but it was not until past the middle of the twentieth century that the development with which we are concerned was

---

many more (see, e.g., the bibliography to Sneath 1962) which in a full reckoning, by a historian of science perhaps, would deserve exploitation. For a short but fundamental statement of the development of taxonomic theory since 1851, see Gilmour (1951).

7. It is only by way of this reference, I am afraid, that I have belatedly picked up the significance of Bambrough's allusion to "botanical taxonomists" in his demonstration that five objects may each have four out of five given features and that the missing feature may be different in each case (Bambrough 1961: 209–10). The example he gives is, with a slight alteration in designations: (1) *A B C D,* (2) *A B C E,* (3) *A B D E,* (4) *A C D E,* (5) *B C D E.*

brought to a point of achievement. Sokal and Sneath write (1963: 13):

> Biologists owe a debt of gratitude to Beckner (1959) for the first clear enunciation known to us of one important concept of natural taxa, a concept which Beckner calls "polytypic."

This concept, which significantly Beckner remarks is "not restricted to biological theory" (Beckner 1959: 21), is contrasted with that of "monotypic" classification, from which we may best approach its meaning. "Monotypic" is a concept "defined by reference to a property which is necessary and sufficient for membership in its extension" (p. 23); i.e., it is equivalent to the traditional common-feature definition of a class. A "polytypic" class, on the other hand, is formally defined (p. 22) as follows:

> A class is ordinarily defined by reference to a set of properties which are both necessary and sufficient (by stipulation) for membership in the class. It is possible, however, to define a group K in terms of a set G of properties $f, f_2, \ldots, f_n$ in a different manner. Suppose we have an aggregation of individuals (we shall not as yet call them a class) such that:
>
> 1. each one possesses a large (but unspecified) number of the properties in G;
> 2. each f in G is possessed by large numbers of these individuals; and
> 3. no f in G is possessed by every individual in the aggregate.
>
> By the terms of 3, no f is necessary for membership in this aggregate; and nothing has been said to warrant or rule out the possibility that some f in G is sufficient for membership in the aggregate.

Beckner goes on to state that a class is polytypic if the first two conditions are met, and "fully" polytypic if condition 3 is also fulfilled. If the number of individuals is "large," all the members of K will resemble one another, though they will not resemble one another in respect to a given f. "If the n [the number] is very large, it would be possible to arrange the members of K along a line in such a way that each individual resembles his nearest neighbors very closely and his furthest neighbors less closely. The members near the extremes would resemble each other hardly at all, e.g., they might have none of the f's in

G in common" (pp. 22–23). In the case of a fully polytypic class, "no f is universally distributed in the class" (p. 24).

A consequence of a polytypic definition of classes is that "there will always be the possibility of borderline cases. . . . Indeed, it is an essential aspect of polytypic classes" (p. 24). Furthermore, Beckner continues (p. 25):

> In a sense, the whole point of polytypic . . . definition is to avoid committing oneself to a necessarily arbitrary delimitation of a class before a theoretically adequate definition can be found. . . . [It] leaves the borderline between K and non-K indeterminate where there is no theoretical reason for drawing the borderline at a particular point. Polytypic concepts are sufficiently justified if only on the grounds of scientific economy: new knowledge can be utilized in applying polytypic concepts without the necessity of modifying their definitions.

As an example of the application of this concept, Beckner later cites "escape reactions," such as heading for the rocks, withdrawing into a shell, running, and so on. It would be very difficult, he says, to attempt to specify a set of characteristics, other than the functional one, which all these responses, and no others, possess. "The class of escape reactions . . . is fully polytypic with respect to those features of behavior observable in the single response, and with respect to movements in relation to any environmental coordinate system, e.g., movements toward or away from particular things" (pp. 122–23).

Here, then, we have a clear formal statement of the taxonomic principle that I was only later to advocate for social anthropology. But the great interest of the conceptual advance thus made is much augmented by a circumstance that I have not mentioned. In his commentary on the concept of a "polytypic" class, Beckner makes a philosophical allusion (p. 23) that presents a new aspect of the convergence of thought that I wish to register:

> Wittgenstein has emphasized the importance that concepts of this logical character assume in ordinary language, especially in that small segment of ordinary language that contains the semantical concepts of "meaning," "referring," "description," etc. He points out that all the members of such classes have a "family resemblance" to one another; he does not suggest a general term for classes of this kind.

Beckner does not cite any particular work by Wittgenstein, though it is most likely to have been the *Philosophical Investigations* (1953). Nor does he say what part, if any, Wittgenstein's conception had in the formulation of his own statement of the principle, and on this point we may be satisfied enough just to note the reference.

The concept of a polytypic class was afterwards taken up by Simpson in his standard work on principles of animal taxonomy (Simpson 1961). There he first expounds the principles of hierarchy and key classification, and then presents by contrast an account of polytypic grouping, described as a method which results in the formation of a taxon in which each member has a majority of the total attributes. "This is a perfectly definite, taxonomically valid and meaningful procedure that involves *no* characters in common . . ." (p. 42). As a hypothetical example of the method, he groups six individuals according to seven attributes (*a*, . . . , *g*) into two species, as in the present table 2.

TABLE 2.   *Polytypic Grouping**

Individuals:	1	2	3	4	5	6
	a	a	b	a	a	b
Attributes:	b	b	c	e	e	f
	c	d	d	f	g	g
	Species I			Species II		

*After Simpson 1961.

The classification is adequate for definition, in that each species has a majority of the attributes. It is adequate also for diagnosis: "In Species I, each member has at least one of the attributes *c*, *d*, neither of which occurs in Species II, and in Species II each member has at least one of the attributes *e*, *f*, *g*, none of which occurs in Species I" (p. 42). Simpson refers in this connection to Beckner's formulation of the principle. In his discussion he adds the significant points, by comparison with traditional taxonomies, that "Species II has no 'differentia' and . . . the two [species], although to us they seem obviously related, have no scholastic or Linnaean 'genus'" (p. 43). Later in his text he gives an account of Beckner's views on polytypy, as demonstration of the fact that "taxa may be recognized and defined by balances or chains of resemblances regardless of characters in common and without having abstractable archetypes" (pp. 94–95). In this phrase there are three terms of special interest: balance, with its statistical implications; chain, with its figurative correspondence with Vygotsky's idiom and its formal parallel with Wittgenstein's overlapping fibers in a rope; and the idea of an abstractable archetype, as the metaphorical paradigm of traditional (monotypic) classification.

The designation of the principle, next, was called into question by Sneath (1962). In considering Beckner's distinction between monotypic and polytypic classification, he raises the objection that these terms have other meanings that are already well established in systematics (p. 291). As better alternatives he proposes "monothetic" and "polythetic" (Gr. *mono*, one; *poly*, many; *thetos*, arrangement). These terms have since been generally adopted in the taxonomies of various sciences (cf. Sokal and Sneath 1963: 13). Sneath glosses the notion of *polythetic* with a reference to phenetic groups which are "composed of organisms with the highest overall similarity, and this means that no single feature is either essential to group membership or is sufficient to make an organism a member of the group" (p. 291). This method of grouping is described as "Adansonian classification" (p. 292), which rightly confirms that the polythetic principle is by no means a modernity in natural science, let alone a novelty contrived in modern social anthropology. But Sneath at the same time makes the significant operational point that "Adanson's ideas could not show their value until the statistical procedures which they required could be handled by modern electronic computing machines . . ." (p. 292). It is only in recent years, some two centuries after Adanson's enunciation of the taxonomic principle, that this has at last been made possible and that the idea of a polythetic class has been applied to full effect. This has been done most notably in numerical taxonomy, where one marked advantage over monothetic classification is, Sneath claims, that "numerical taxonomies . . . are unlikely to need radical revision, since, being polythetic, they are not liable to the catastrophic effects of finding that a character of presumed crucial importance is after all useless" (p. 325).

Yet it was soon shown that a polythetic classification did not necessarily compel the revision of a monothetic classification of the same materials. Lockhart and Hartman, in a report on a test of method in quantitative bacterial taxonomy (1963), begin by defining polythetic groups as those in which "no property is necessarily possessed by all individuals in the group, and no organism necessarily has all the properties generally characteristic of members of its group." In consequence, "any given organism may score mathematically as being equally similar (though in different respects) to two or more other individuals which, in turn, are quite dissimilar to each other" (p. 68). Thus polythetic groups are "not mutually exclusive" and are "not theoretically analogous to the hierarchical taxa of present classifications" (p. 69). Nevertheless, a computerized analysis of 50 bacterial strains, classed by 60 properties, first by polythetic grouping and then again by a monothetic grouping, produced little difference in the results. "Rather curiously, our monothetic groupings were nearly identical to those obtained on a polythetic basis. It is conceivable, though it seems quite unlikely, that this apparent identity is coincidental and unique to the material used. . . ." These findings, which are illustrated by two remarkably congruent tree-diagrams (pp. 71, fig. 1, and 73, fig. 2), suggest to the experimenters that "the polythetic groups of organisms occurring in nature have a monothetic core of common properties" (p. 76). This would entail, in their opinion, the allaying of fears that "it might one day be necessary to abolish present classifications and to replace them with taxonomies and diagnostic schemes which somehow would take into account the multidimensional polythetic arrangements of organisms which actually seem to occur in nature" (p. 77).

Within the same year as this critical test, Sokal and Sneath (1963) published a definitive statement of the principle of polythetic classification, in their comparative and methodological textbook on numerical taxonomy. They first establish the principle of monothetic classification: "The ruling idea of monothetic groups is that they are formed by rigid and successive logical divisions so that the possession of a unique set of features is both sufficient and necessary for membership in the group thus defined"; they are called monothetic, it is explained, "because the defining features are unique" (p. 13). "A polythetic arrangement, on the other hand, places together organisms that have the greatest number of shared features, and no single feature is either essential to group membership or is sufficient to make an organism a member of the group" (p. 14). For a formal expression of this concept, the authors declare that they cannot do better than to quote Beckner's definition (see above). They add, however, the operational qualification that natural taxa are usually not fully polythetic, since one can usually find some characters common to all members of a taxon: "It is possible that they are never fully polythetic, because there may be some characters (or genes) which are identical in all members of a given taxon." All the same, they conclude: "We must regard a taxon operationally as being possibly fully polythetic, since we cannot be sure that we have observed any characters that are common to all members" (p. 14).

Sokal and Sneath supply as a formal illustration the scheme reproduced here as table 3. Here, "the class of 1 + 2 + 3 + 4 is polythetic (and in this instance is also fully polythetic, since no one character is found in all the four individuals). . . . Individuals 5 and 6, however, form a monothetic group" (p. 15).

TABLE 3.   *Polythetic Arrangement**

	Individuals					
	1	2	3	4	5	6
	A		A	A		
	B	B	B			
Characters	C	C		C		
		D	D	D		
					F	F
					G	G
					H	H

*After Sokal and Sneath 1963.

One of the difficulties of Beckner's definition, Sokal and Sneath add, is that in natural taxa we do commonly find features that are not possessed by large numbers of members of the class. "Furthermore, we cannot test whether any given f [feature, character] is possessed by large numbers of the class before we have made the class, and therefore we cannot decide whether to admit this f into the set G'' (p. 15). But this difficulty, they state, can be avoided by defining membership of the class in terms of common (or shared) attributes. Further on the score of method, they add that polythetic groups can themselves be arranged polythetically to give higher polythetic groups, "as is done in building a hierarchy in a natural system." As for the consequences of adopting a polythetic taxonomic principle, the advantages of polythetic groups are that "they are 'natural,' have a high content of information, and are useful for many purposes"; the disadvantages are that "they may partly overlap one another (so that hierarchies and keys are less easy to make than with monothetic groups) and that they are not perfectly suited for any single purpose" (p. 15). By way of a general reminder, the authors later stress that "when . . . groups are polythetic ones, we must bear in mind that it is never certain, but only more or less probable, that a member possesses any given feature" (p. 171).

Finally, for the purposes of the present survey, we may notice the cast more recently given to the principle by Mayr, citing Beckner and Simpson, in his textbook on systematic zoology (1969). "Taxa characterized by a set of characters of which each member has a majority are called *polythetic* taxa. . . . No single feature is essential for membership in a polythetically defined taxon nor is any feature sufficient for such membership" (p. 83). In fact, as he continues, many zoological taxa "are based on a combination of characters, and frequently not a single one of these characters is present in all members of the taxon . . ." (p. 88). Mayr's glossary defines "polythetic" as: "of taxa, in which each member has a majority of a set of characters" (p. 409). In this formulation of the principle, then, the stress is laid on the possession by each member of a class of a simple majority of the defining features, without, that is, the assignment of decisive weight to any of them.

IV

We see therefore that polythetic classification, far from being a unique innovation in current social anthropology, is a recognized taxonomic principle in a range of natural sciences.

In the adoption of this principle, two circumstances appear to have been crucial: the promulgation in 1859 of the theory of evolution, when natural history challenged the logicians' method of classifying species; and the development, a hundred years later, of electronic computers, which made it practicable at last to put into scientific effect the precept enunciated by Adanson in 1763. In the outcome, polythetic classification became employed in sciences so various as botany, zoology, biology, and bacteriology, and it carried fundamental implications for each. The empirical application of the method, and the elaboration of the principle, were in the main the work of a period as recent as the 1960s.[8] In the formal expression of the concept of polythetic classification, an especially significant point is the allusion by Beckner to Wittgenstein's emphasis on concepts of the same logical character in ordinary language. This neatly clinches a noteworthy convergence in the history of ideas, and on this score the main part of my case is made.

Let us now examine some of the consequences for social anthropology that may be derived from this better appreciation of the polythetic principle. This will have to be done in fairly general terms, without a conclusive empirical demonstration of each point of argument, but the practical relevance of the principle to the comparative study of social facts should be evident enough.

I have already entered the caution that, in the study of descent systems in particular, a direct consequence of polythetic classification in anthropology is that "comparison becomes far more difficult, and on any large and detailed scale perhaps impracticable" (Needham 1974a: 50). In the study of relationship terminologies, similarly, the consequence is that "while analysis is made more exact, comparison is made more intricate and dif-

---

8. To stress the temporal continuity in the convergence that I am demonstrating, it may be noted that Mayr's book on systematic zoology was published in 1969 and that the idea of classifying by "serial likenesses" in social anthropology was proposed to the A.S.A. conference on kinship and marriage in the spring of 1970 (see Needham 1971a). The interval between even Beckner's monograph and the delivery of the conference paper was eleven years.

ficult" (p. 60). More widely, the consequence is the realization that the social phenomena classed as "kinship," etc., "do not in all cases possess any specific features such as could justify the formulation of general propositions about them" (p. 69); "there may . . . be something in common, under each general term [e.g., 'marriage,' 'incest'], but not necessarily a definite set of characteristic, specific, or essential features" (p. 70). Operationally, this means also that attempts to find statistical correlations are seriously affected, for these cannot be established without a precisely specified typology of phenomena; but if it is conceded that the social facts in question do not necessarily compose a conventional class of a homogeneous kind, under each type, "but may instead exhibit an immense array of serial and more complex resemblances," then the grounds for this method of comparison and explanation are removed (p. 70).

By this account of the matter, the first consequence of the adoption of polythetic classification in social anthropology is that comparative studies, whether morphological or functional or statistical, are rendered more daunting and perhaps even unfeasible. Yet polythetic classes are likely to accommodate better than monothetic the variegation of social phenomena: they have, as Sokal and Sneath put it, a high content of information, and they carry less risk of an arbitrary exclusion of significant features. In other words, it could be said, the polythetic principle is truer to the ethnographic materials. If this is so, then the increased difficulty of comparison is a price that simply has to be paid—if there is some way of meeting it.

This is a point on which social anthropology may have much to learn, by way of explicit method, from the sciences in which the polythetic principle has already been well recognized. We may find special instruction, in the first place, under the aspect of evolution, since this is so basic to the sciences whose examples we have just considered. I shall therefore deal in some detail with the methodological implications of studying evolutionary materials.

One very striking difference between the materials that a natural scientist has to classify and those which are the concern of the social anthropologist is the presence or absence of evolutionary connections. It is the factor of common but divergent descent that permits a zoological taxon, for instance, in which not a single one of the definitive characters is present in all members of the taxon (Mayr 1969: 88). This is an acceptable grouping of natural entities, whether animals or plants or bacteria, because lineages of descent can be demonstrated; and this can be done even if intermittently the original characters have changed or fallen away, under the pressures of adaptation, until the descendants bear little resemblance either to one another or to their common ancestor. It is doubtless the factor of phylogenetic descent that encourages some among even the most modern taxonomists to speak of "natural" groupings of organisms, and to distinguish such aggregates from "artificial" groupings. I suppose, too, that it is phylogeny that largely accounts for what Lockhart and Hartman isolate as the "monothetic core of common properties" in polythetic classes of bacterial strains (1963: 76). We might therefore sum up the situation in the natural sciences by saying that, as instrinsic features of the materials under comparison, common descent composes natural classes while natural selection variegates the members of such classes.

In the comparative study of social facts, however, these conditions do not obtain. It is not without good reason that social anthropologists today pay little if any attention to evolutionary connections among social forms, to the extent indeed that it is hard to find a modern instance of such an approach.[9] There is, though, one study of the kind, carried out on a scale comparable with that of zoology or bacteriology, in Murdock's *Social Structure* (1949). In his chapter on "Evolution of Social Organization," Murdock propounds five generalizations about "the normal order of change among the principal elements of social organization" (pp. 221–22), and he applies these generalizations to his sample of 250 societies. He distinguishes eleven "major types of social structure" defined by rules of descent and types of cousin terminology (p. 224), and these major types are subdivided by rules of marital residence into a total of 47 "structural subtypes" (pp. 324–26). A technique for the reconstruction of antecedent forms is presented (pp. 326–31), and

9. Evolutionist explanations, such as were common in the last century, are sharply to be contrasted with historical (or "diffusionist") reconstructions, the feasibility of which is another matter altogether (Needham 1970b: lxv–lxxxi).

each of the societies in the sample is then assigned to a subtype and thence derived from a series of prior subtypes (pp. 332–34). The enterprise is painstaking and detailed, and consistently worked out; but it does not serve to establish a correspondence between natural evolution and social evolution such as could now permit the application in social anthropology of the taxonomic procedures that we have observed in zoology and other natural sciences. I do not refer to defects in the stipulation of those features (e.g., "cousin terminology") by which the types and subtypes are distinguished, for in principle these could be amended and the evolutionist method remain feasible. What is a graver criticism is that Murdock's evolutionary scheme permits the establishment of such a proliferation of transitional connections, many of which are reversible, as to prompt doubt concerning its general validity (Needham 1962: 173–80). Certainly there is extraordinarily little resemblance between the "evolutionary" connections among a set of eight societies that I selected from the sample (p. 178, fig. 1) and the phylogenetic trees that can be established in the natural sciences. Another source of doubt concerning the scheme is that by the introduction of an additional discriminating feature (viz., prescriptive alliance), which Murdock did not take into account, it can be shown that a number of societies in the sample can be placed in quite different relative positions from those which they occupy on Murdock's premises (p. 175). This implies that sociological taxonomy is highly artificial (as in fact it is), and such that the evolution of institutions may be disparately inferred according to the typological features that are more or less arbitrarily decided upon. There is no sign, at any rate, of a correspondence with quasi-natural monothetic clusters of features. Rather, it is the construction of a sociological typology of cultural particulars by monothetic criteria, and these too of the most slender and unreliable kind, that is the chief and irremediable weakness of this evolutionary scheme.

Now Murdock's venture is admirable in its own terms, and in some ways the most impressive evolutionary undertaking in anthropology, but as soon as the polythetic character of the resemblances in question is recognized, the grounds of the taxonomy on which the study rests are quite removed. For example, the Araucanians, Iatmul, Miwok, Nyoro, and Thado are all typed together as "normal Omaha" (Murdock 1949: 240), on the basis of patrilineal descent and the terms for cross-cousins (pp. 239, 244), whereas the quickest survey of the ethnographic evidence on each society shows that their forms of social classification, not to mention their institutions, are systematically disparate (Needham 1974a: 51; cf. Lowie 1917, chap. 5). The concept of "descent," moreover, is itself polythetic (Needham 1974a: 47), and so also are the conventional types by which relationship terminologies are classed as "Omaha," etc. (pp. 47–54, 59–60). When, therefore, the descent systems under comparison are analyzed by polythetic criteria, instead of being typed by a few monothetic features (p. 60), the presumed resemblances are reduced or abolished; the comparison is vitiated, and the attempt to work out evolutionary interconnections is thereby doomed to failure.

This particular outcome does not mean, however, that no form of evolutionary reconstruction is possible in the comparative study of institutions. I have suggested indeed that prescriptive systems tend to change in regular ways (Needham 1967: 46; cf. 1980c), and it seems quite feasible to establish regular transformations which bear a resemblance to evolutionary modifications in natural species. Admitted, this line of inquiry has not yet been taken very far, and the series of transformations so far proposed is very tentative, but there remains nevertheless some empirical justification for speaking in these regards of the "evolution" of social forms. Yet it does not at all follow that the taxonomic methods of the natural sciences, e.g., zoology, can therefore be directly taken over in social anthropology. The crucial feature that is missing from the prescriptive systems compared is still that of common origin. If there is a structural correspondence between the terminologies of the Tibeto-Burman Garo of northeastern India and the Malayo-Polynesian Manggarai of eastern Indonesia (Needham 1966: 155, table 4), this is not because these societies are descended from a common ancestral society. Still less is such an evolutionary connection the explanation of the resemblances among other two-line terminologies in Australia, southern India, and the tropical forests of South America. In fact, the type of evolution at work in these instances is no more than formally similar to that of natural species.

The outcome, in any case, is certainly not much like that in zoology, for example, for whereas a zoological taxon may comprise very dissimilar individuals, the anthropological taxon of prescriptive systems of social classification is so defined that the individual terminologies shall be distinctly alike. The points of resemblance are not cultural particulars but analytical abstractions; prescriptive terminologies are thereby classed monothetically. The resultant class of social facts, defined by a common feature (the invariant relation articulating the categories) is artificial. All the same, the label "prescriptive" does not denote a class of societies to which the law of substitution can be applied, for in prescriptive alliance systems (conceived as empirical forms of social life) the categories, marriage rules, groupings, and modes of social action are independent variables (Needham 1973b: 174). In other words, whereas prescriptive terminologies are defined monothetically, the class of "natural" societies characterized by jural categories of this type is polythetic. It may be possible to list certain characteristic features of such societies (Needham 1970a: 257; cf. Korn 1973: 100–04), but the societies do not compose a class such that "whatever we know of one object . . . we also know of the other objects."

Even in this quite promising line of "evolutionary" investigation, therefore, the parallels with a natural science are superficial. An abstract stipulation occupies the place of a common ancestor; the similarity of social forms has nothing to do with common descent; and the postulated transformations of prescriptive systems appear causally different (Needham 1967: 47) from the evolution of natural species. These essential contrasts make it hardly likely that the methods of the natural sciences, in coping with polythetic classes of evolutionary materials, should be directly transferable into social anthropology.

Nevertheless, the polythetic approach in the natural sciences can still seem to offer the promise of a method that might be adapted to the concerns of social anthropology. This is in the practice of numerical taxonomy, and particularly in the method of "cluster analysis." We have seen that Adanson's principles demanded statistical operations that were not practicable until the development of electronic computers. Likewise, an obvious difficulty in the large-scale comparison of social facts has long been that the great numbers of variables involved made the task prohibitive; and this source of difficulty is very much magnified when the evidences compared are defined polythetically. But in quantitative bacteriology, Lockhart and Hartman have managed to compare 50 organisms by reference to 60 properties such as production of indole, colony morphology, and tolerance to sodium chloride (1963: 69–70). This is impressive as it stands, but the researchers add that a certain computer could be programmed to handle as many as 1,500 bacterial strains, each one scored for 200 properties. With figures so high as these we approach the sort of computation that presumably would be called for in the comparison of social facts. If therefore the polythetic classification of bacteria can be coped with by an appropriate computer program, perhaps the taxonomic principle in itself need no longer be seen as an obstacle to effective large-scale comparison in social anthropology.

Unfortunately, Lockhart and Hartman's experiment does not make the case, for by their method the groups "are *made* monothetic by discarding all characters which vary within them" (Sokal and Sneath 1963: 188; emphasis added). Without going into the technicalities of this procedure so far as bacteriology is concerned, we can at once see that it introduces an arbitrary factor such as in social anthropology we seek to obviate precisely by the adoption of polythetic classification. In the study of social facts, what is needed is not a convenient technique for cutting down the number of variables, but a means of accommodating as many as possible. The method of cluster analysis can be justified in the case of Lockhart and Hartman's experiment by the result that the polythetic groupings were nearly identical with the monothetic; but this can be explained by their suggestion that polythetic groups of organisms occurring in nature have a monothetic core of common properties, and such phylogenetic properties do not characterize social facts. We can well follow the bacteriologists in their application of the taxonomic principle, but the evolutionary nature of their materials introduces a difference of kind which prevents us from directly taking up their statistical methods.

There are in any case other reasons for not turning to cluster analysis for the purposes of anthropological comparison. Quantitative methods in taxonomy are "based on the principles that classification is a measure of overall similarity among organisms and that all properties of organisms are potentially of equal value in creating taxa, so that no *a priori* assumptions need to be made of the relative importance of particular features" (Lockhart and Hartman 1963: 68). Simpson writes of "balances of resemblances regardless of characters in common" (1961: 94–95); Sneath refers to the "highest overall similarity" (1962: 291); Mayr defines a polythetic taxon as one in which each member possesses simply a "majority" of a set of characters (1969: 83, 409). This approach entails a number of difficulties when we conjecture its possible use in the comparative study of social facts.

To begin with, "it is reasonable to ask whether a definition of a polytypic concept is after all a *definition*, since it is certainly imprecise" (Beckner 1959: 24). In Beckner's formulation, as he points out, "the vague term 'large number' occurs twice in the definition" (p. 24), and there is no rule of method for deciding in general, or for any given context, what is to count as a large number. The same kind of uncertainty attaches to the phrases "overall similarity," "balances of resemblances," and a "majority" of characters. No matter how the definition of a polythetic class is expressed, the difficulty is to know where to draw the line. This problem is not resolved by the admission that "there will always be the possibility of borderline cases" (Beckner 1959: 24), for the location of the border is itself a function of the degree of numerical preponderance that is thought sufficient, and this in principle is always contestable. In any event, the consequence is that "it is never certain, but only more or less probable, that a member [of a polythetic group] possesses any given feature" (Sokal and Sneath 1963: 171). A numerical taxonomy, therefore, leaves the social anthropologist in much the same definitional quandary as when he is faced with the question of what is to count as an instance of a given institution. For example, in a reassessment of the concept of marriage, Rivière has concluded that "we mislead ourselves by describing with a single term relationships which in different societies have no single feature in common other than that they are concerned with the conceptual roles of male and female" (Rivière 1971: 70). That is to say, "marriage" is what Wittgenstein called an "odd-job" word (cf. Needham 1972: 124–25; 1974a: 44); or, as we might better say now, it is a polythetic concept. No statistical method or computer program can decide what is to count as marriage. Concomitantly, to identify a particular social relationship as an instance of marriage, by a polythetic definition, will not demand the presence of any specific feature by which it is possible to decide that it is to count as marriage.

On the other hand, a polythetic conception of marriage, as determined by overall similarities, still does not preclude the risk of leaving out of account some feature that is regarded indigenously as essential to the relationship, or one that might be more relevant to the purpose of the comparison. This is especially likely when what is in question is an evaluation placed on the relationship, e.g., a moral repugnance for divorce, which happens not to be present in a majority of the cases classed together. In general, indeed, human affairs are semantically so very complex that it must be difficult in the extreme (if it is even conceivable) either to stipulate significance as a polythetic feature or to assess the degree of similarity among the meanings or values attached by different civilizations to any kind of institution that is the subject of a comparative proposition.

Another difficulty is exposed by the possibility in natural science of determining features by difference, e.g., by whether or not an organism produces indole or ferments lactose. "There may be endless arguments as to whether two organisms are similar in that neither ferments lactose, but everyone agrees that they are different if one does so and the other does not" (Lockhart and Hartman 1963: 70, 77). But what in the study of human affairs would permit this decisive treatment? At the level of social facts, as distinguished from analytical abstractions, it will be usually dubitable and often impossible to determine such absolute contrasts. For that matter, though, it is not always so easy even in the natural sciences to determine features so unambiguously. In bacteriology, for example, the Lockhart and Hartman experiment classified organisms by features such as colony morphology and tolerance to NaCl, but these involved relative discriminations and hence a degree of arbitrariness. Three forms of colony were scored, viz., punctiform, circular, and "irregular"; and tolerance to NaCl was scored at four levels as sensitive, weakly resistant, moderately resistant, and strongly resistant (Lockhart and Hartman 1963: 70). These morphological and scalar discriminations are uncertain and at least imprecise, and some are no more dependable than are the criteria often employed in social anthropology. Moreover, even on the assumption that features can be well determined, numerical taxonomy has a disconcerting consequence that must limit or even rule out its application in social anthropology. In a polythetic group, as will be remembered, no property is necessarily possessed by all members in a group, and no individual necessarily has all the properties generally characteristic of members of its group. Hence, in bacteriology, "any given organism may score mathematically as being equally similar (though in different respects) to two or more other individuals which, in turn, are quite dissimilar to each other" (Lockhart and Hartman 1963: 68). Clearly this is an outcome that would be quite unacceptable in the study of social facts.

Underlying all these difficulties that lie in the way of adapting numerical taxonomy to the polythetic concepts that are nonetheless called for in social anthropology, there is a more fundamental obstacle. In the natural sciences the features by which polythetic classes are defined have generally a real, distinct, and independent character, and they can be clearly stipulated in advance. Such features are, in zoology, skeletal structure (a definite number of bones functionally arranged in a certain order); in botany, roots, leaves, pistils; in bacteriology, chemical elements, compounds, and their reactions; in many sciences, at a deep level of analysis, molecular structures and the particles of which these are composed. This is a rather rough and ready characterization, of course, and it becomes less appropriate when ultimately the character of "fundamental" particles comes into question, but it serves all the same to make a crucial contrast between the natural sciences and social anthropology. In what has been presented here as the most relevant example of taxonomic method, namely quantitative bacteriology, the researchers are in no doubt concerning what is or is not lactose or about whether it is or is not present: it can be exactly defined in advance, and its chemical properties and reactions are known or testable. This kind of certainty about the materials under study (whatever ambiguity may attend the discrimination of forms or the assessment of degrees of resistance, etc.) permits the method of classification by differences: a definite feature can be definitely determined as either present or absent.

But in the realm of social facts this aspect of polythetic classification is hardly to be found. A main reason is that in social
anthropology the determination of the constituent features of
a polythetic class cannot be carried out by reference to discrete
empirical particulars, but entails instead a reliance on further
features of the same character which themselves are likewise
polythetic. In social life, that is, there are no established phenomena, in the form of isolable social facts, for instance, which
correspond to the elements and particles in nature. The disparity between the natural sciences and social anthropology, in
taxonomic method as in much else, reflects a contrast of kind
between natural entities and social facts. This contrast is the
most marked when the materials for an anthropological classification are collective representations. More generally, in any
case, there is no reason that a classificatory technique that is
appropriate to one kind of evidence should be applicable to another, and all the less is this so when the evidences in question
are contrasted as physical and ideational.

V

In this essay I have set on record what strikes me
as a remarkable convergence in the history of ideas, and I have
briefly considered certain methodological consequences of the
discovery that classification by sporadic likenesses is already
practiced in certain natural sciences.

But the results are not encouraging, for whereas the polythetic methods developed in those sciences are much in advance of those suggested for social anthropology (Needham
1971b), both in their formal expression and in their empirical
application, there are serious objections to the adoption of
such methods into the study of social facts. This setback is not
the fault of the taxonomic principle, but has to do with the essential disparity between natural phenomena and collective
representations. In social anthropology the idea of a polythetic
class does indeed proffer a great rectification of thought and a
newly panoptic perspective on human affairs. But at the same
time this vantage offers an austere prospect, for amid so many
uncertainties as we have glimpsed, the one sure outcome is

that comparative studies, if they are carried out in empirical terms, will become irresolubly complicated. Polythetic classification therefore tends, by this view, to confirm Evans-Pritchard's aphorism: "There's only one method in social anthropology, the comparative method—and that's impossible."[10]

There is a great deal to be said perfectly seriously in support of this quip, but the issue is not quite so decided as it implies. Comparative studies are likely to be defective and unproductive so long as they continue to be carried out within conventional, i.e., monothetic, taxonomies and by reliance on substantive paradigms (Needham 1974a: 53, 60). But comparison stands a better and quite different chance of success if it is conducted in formal terms. There have been prominent examples of this kind of approach under the rubrics of "elementary forms" and "elementary structures," and it could yet be pursued further by the formulation of a "vocabulary of analytical concepts that were appropriate to the phenomena under consideration but would not be merely derived from them" (Needham 1974a: 16). Not only are the terms of common English, such as "kinship" or "marriage," worse than unserviceable in this enterprise, but so also are the quasi-technical terms of anthropology such as "unilineal," "Omaha," "patrilocal," and so on. As will be appreciated by this point, a characteristic flaw in such terms is that they have been presumed to denote monothetic classes of social facts, and thus to permit the operation of substitution; whereas in fact their reference is polythetic, so that comparative propositions are perpetually undermined by the awkwardness that the absence of common features precludes the possibility of substitution. But this objection does not attach to relational concepts such as "symmetry," "alternation," "transitivity," "complementarity," etc., or to analysis by reference to logical possibilities (Needham 1974a: 16, 47; 1974b: 39). Here we have formal properties which can be defined in purely formal terms, e.g., in the notation of symbolic logic, without reference to any classes of entities, however the classes may be composed, or to the characteristic empirical features of their members.

---

10. He said it to me, and doubtless to others as well, but so far as I know it does not appear in any of his published writings; cf. Evans-Pritchard (1965).

Formal analysis of this kind is not merely a methodological postulation, nor just a critical technique in the undermining of received categories of anthropological discourse. In practice, it accurately accommodates social facts, it facilitates systematic analysis, and it makes possible the effective prosecution of comparative studies. In each of these regards this radical style of abstraction evades the disastrous deception of reificatory and monothetic taxonomies; and it renders unnecessary a desperate contention with the alternative hazards of trying to theorize taxonomically about classes of facts that in empirical terms are polythetic.

The formal terms mentioned above, together with others that propose themselves as appropriate to any relational inquiry, have been arrived at largely inductively through their application in the empirical analysis of various types of society. They also happen to accord, however, with certain philosophical conclusions arrived at by Campbell in his study of family resemblance predicates (1965). He argues that, for any given linguistic context, not every predicate can be of this kind; so "it follows that the notion of family resemblance cannot of itself solve the problem of universals"; there must be, for any such context, what he calls "basic predicates" (p. 243). In social anthropology there have been repeated attempts to establish ultimate theoretical notions, relating to social life and collective representations, in morphological or functional or structural terms, or else by resort to universal needs or motives or mental capacities. These ventures have not proved successful, but the relational concepts that I have adduced are by contrast well adapted, I suggest, to the analytical task of taking each case as it comes and to the comparative task of formulating propositions about classes of social facts. There is pragmatic reason to think that such purely formal terms may provide, for certain purposes, the basic predicates that are called for in social anthropology.

The present article also accords in another important regard with the conclusion to Campbell's analysis. Where we wish to make generalizations in the confidence that they admit no exceptions, he writes, defined terms are to be preferred, other things being equal, to family resemblance terms: "We should not rest content until family resemblance predicates, admittedly intelligible, have been banished from our sciences" (Campbell 1965: 244). This is exactly the aim of my "Remarks," to which this essay forms a sequel, though I do not think that such (polythetic) predicates can ever be eliminated from practical description in the field or from academic discourse about ethnographic reports. Where they can deliberately be dispensed with is in the contrivance of a formal theoretical terminology. Studies of prescriptive alliance systems show particularly well that a definition which is designedly monothetic, in combination with analytical concepts which are necessarily of the same character, can advantageously be applied to classes of social facts that are extensively polythetic.

There is a great deal more that could be written about these fundamental topics, and what I have written here is no more than groundwork for the formulation of theoretical issues, both formal and analytical. Also, if the appreciation of polythetic classification is to have its proper effect, it is desirable that more detailed empirical demonstrations should be fully set out. For the present, though, I conclude with two observations that indicate something of the relevance of this theme to ultimate concerns in the understanding of human affairs.

The first is to take up the point that "our analysis may be guided by the same logical constraints as must have been effective in producing the systems that we study" (Needham 1974a: 36, 48; 1974b: 39–40). This is particularly clear in the comparative study of descent systems, terminologies, and certain rules of marriage, but there is a wider implication. If we consider that formal terms such as "symmetry" or "transitivity" are not peculiar to a particular linguistic and intellectual tradition, but denote properties which must be discriminable (either conceptually or in social practice) by any cultural system of thought, then it follows that the terms are intrinsically appropriate to the study of exotic collective representations. Alternatively, a more speculative notion is that the formal terms denote mental proclivities and constraints which are universal to mankind in the fabrication (deliberate or not) of categories and articulatory relationships. According to either of these conjectures, the kind of theoretical terminology to which I have referred would thus naturally qualify as basic predicates.

The other comment, linked to the former, is to stress the conception of social anthropology that inspires the present essay. The approach is guided by the ambition, inherited from the *Année sociologique* school and directly traceable back to the Enlightenment, to determine what can be called the primary factors of experience by the comparative study of social facts. The facts at issue in the "Remarks" and in my analysis of belief were words, viz., verbal concepts framed by cultural traditions in the classification of the world, whether this was the ordering of social life or the ascription of capacities to the human mind. In the analysis of such collective representations, whether they were the quasi-technical generalizations of social anthropologists or the ordinary resources of everyday language, it was found that certain deep miscomprehensions were ultimately the results of the traditional assumption that classificatory concepts were necessarily composed about common definitive features, i.e., that they were monothetic. In each case, however, it was shown that the words in question actually denoted classes composed by family or sporadic resemblances, i.e., that they were polythetic.

Now the outcome of analyses of this kind should not be seen as merely a local or technical rectification of European academic argument, but as pointing to a general hazard of language which presumably afflicts men in any tradition when they classify their fellows and their nature. "In seeking to translate alien concepts . . . we have to appreciate that the foreign words in question are themselves words that may be in the same state as our own," so that the speakers of another language, constrained through it by their own collective representations, "must be assumed to be the victims of just such linguistic defects, traps, and diversions as are we ourselves when we formulate our own thoughts" (Needham 1972: 233). Thus the realization of the confusions brought about in social anthropology by stock classificatory terms may serve to prepare our understanding in coming to terms with alien concepts which, in a fashion that is similarly unrecognized by those whose modes of thought we want to comprehend, are also polythetic.

The concept of cause . . . signifies something different
from "that which happens," and is not therefore
contained in this latter representation.

Kant

4

Skulls and Causality

I

The question that I take up here concerns the influence of our conception of causality on the interpretation of head-hunting. The empirical evidence from which I begin is taken from Izikowitz's ethnographic accounts of the Lamet of Laos, Indochina.* The Lamet themselves are not in fact head-hunters (Izikowitz 1941: 27; 1951: 25), but they have a mystical concept which leads directly to the problem that I wish to consider.

The concept is that which is denoted in Lamet by the word *klpu*. It is introduced to us in connection with what Izikowitz at first refers to as the "soul" of rice. Rice "possesses" a *klpu*, but other plants such as root-crops, millet, eleusine, coix, sorghum, and several kinds of spices do not. The only other thing that has *klpu* is the human being, and in this case "two of them,

---

* This essay was written, in March 1974, as a contribution to a proposed Festschrift that was to be dedicated to Professor Dr. K. G. Izikowitz, of the University of Göteborg, on the occasion of his seventieth birthday.

one situated in the head and the other in the knees, each ruling the upper and lower parts of the body respectively, and finding their boundary line in the region of the navel" (1941: 7). More particularly described, it is "concentrated" in the head and the knees, but "it is to be found all over the body" (p. 28).

The *klpu* is not a "spirit" (*mbrōng*): "it is something that gives life to the body and to the rice . . ." (p. 9). It has no form, and there is no connection whatever between the *klpu* of a man and the specter of a dead human being (p. 10), nor has it anything to do with personality (p. 24). The *klpu* cannot be powerful or weak, and one cannot have more or less of it; nor can it be transmitted to others (p.10). But it has an "unsteady and fleeting nature," and special acts have to be performed to keep it in place. After a burial the *klpu* of the deceased rests in the grave, fastened by stones, while the spirit goes away to the village of the dead (p.14).

As for the *klpu* of rice, this also has an unsteady attachment to its habitat: it is "a sort of living quality" which can easily leave the plant (p. 15), and the Lamet explain that they desire the *klpu* of rice from neighboring lands to join the *klpu* in their own fields and thus produce a bigger crop and also make the rice last longer (p. 19). At harvesting, some of the *klpu* lies in the grain, but part is left in the harvested area of the field, from which it might "run away," so it has to be "guided" over to an unharvested area. It does not walk like an ordinary being, but "just drifts in an indefinable way." Since it cannot cross open spaces alone, it is conducted over paths by charred branches decorated with bouquets of flowers (p. 21). It is afraid of loud noises, and a gunshot can cause it to leave the rice fields (p. 22). Just as it can be dispersed, so it can also be concentrated: it is "collected" with the harvested rice and is secured in the granary, where it is represented by a sheaf taken from the last place in the field that was reaped, i.e., the ultimate spot where the *klpu* had taken refuge. As at a grave, a stone is placed in the final storing-place of the rice in order to "fasten" the *klpu*, which otherwise might follow the piles of the granary and run down into the earth. Bamboos are tied around the piles, also, as a boundary line forbidding the *klpu* to pass (p. 23). The rice crop depends on the "quantity" of the rice-*klpu* (p. 24).

The "two kinds" of *klpu*—that of human beings and that of

rice—thus have "some characteristics in common, and both function as a kind of living entity that can be tied or fastened," failing which they are likely to disappear. They are different, however, in that the *klpu* of rice can be increased in quantity and that it can be guided (p. 24).

Izikowitz concludes that "the *klpu*-conception of the Lamet is to some extent related to the idea of 'soul-substance' in Indonesia" (p. 27; cf. p. 9).

Since he suspected that in animals at any rate the skull might in some way be regarded as "a center of power," he cross-examined the Lamet on this point, but could get no information (1951: 334). Nevertheless, on this particular question he does not consider it improbable that the skull is a center of power. He is encouraged in this idea by the fact that the Puli-Akha, a neighboring tribe, have the idea of a head-soul which represents the sexual power of a man and is his center of strength, such that if the plait of hair in which it is concentrated were to be cut from the man's head he would "wither away and die" (p. 94). In support of this supposition about the Lamet, Izikowitz further adduces the case of the Wa, who are "closely related to the Lamet," among whom human heads are used in sacrifices made to the spirits of the fields in order to get better harvests (p. 334). This idea, he continues, seems to be the basis for the use of heads as trophies elsewhere in southeast Asia, among the Dayak of Borneo and the Naga of the Indo-Burma border, and in other parts of the world as well. "The head becomes the site of a kind of life 'energy,' which the owner of the head can make use of" (pp. 334–35).

II

The idea of "soul-substance," or what alternatively Izikowitz calls "life-energy," can be traced back to the great Dutch missionary ethnographer A. C. Kruyt.

In his work on animism in the Indonesian archipelago (1906), Kruyt writes of a general notion among the peoples of that part of the world that the entirety of nature is animated by an all-pervading "soul," a fine ethereal stuff which imparts life.

He had previously referred to this life-giving matter as *levens-fluïde,* "life-fluid," but P. D. Chantepie de La Saussaye proposed that the indigenous idea would be much better expressed by *zielestof,* "soul-substance" (Kruyt 1906: 2).

Primitive man is said by Kruyt to regard the head as preeminently the site of soul-substance in the human being. Accordingly, "the great importance of the head as container of soul-substance [*zielestof*] emerges best in the custom of head-hunting" (p. 17). The motive behind this custom, it is alleged, is to make use of the enemy's soul-substance by taking away his head.

An excellent illustration of this belief is provided by a story from the Kĕnja (Kenyah) of the Baram river, in the interior of Sarawak, Borneo, as recounted by Kükenthal. This relates that formerly men used to cut off only the hair of their fallen enemies, and that they used this to decorate their shields. A tortoise advised the great chief Tokong to cut off instead the heads of his enemies, and when this counsel was repeated in a dream Tokong acted upon it. It was then observed that those who carried the heads were far ahead of the others, and that they experienced no fatigue. Also, the river began to flow upstream and took them swiftly home. The rice, too, grew under their very eyes, and the old and sick in the village were made healthy. These marvels were brought about by the human heads (Kruyt 1906: 18; citing Kükenthal 1896: 280–81).[1] In other words, explains Kruyt, the soul-substance of the human heads is communicated to everything and everybody.

Apart from such tales, however, the chief ground for Kruyt's opinion on this score was his long residence among the Toradja of Celebes, who had been head-hunters. Kruyt formed the idea that the eastern Toradja conceive their welfare as depen-

---

1. In a Sebop version reported by Hose, it was a frog which told Tokong to cut through the neck. The frog assured him that the taking of heads would bring prosperity of every kind, and Tokong acted on this advice. As the successful war-party returned home with the heads, and passed through the rice fields, the rice grew very rapidly: "as they entered the fields the *padi* was only up to their knees, but before they had passed through it was full-grown with full ears." When they approached the house, their relatives came out to meet them, rejoicing at various pieces of good fortune that had befallen them (Hose and McDougall 1912, II: 138–39; cf. Furness 1902: 59–61).

dent on a supply of *tanoana*, which he glosses as *levenskracht*, "life-force." Women, as priestesses, procure this from heaven; men, as warriors, do so by taking heads (Adriani and Kruyt 1951, II: 77). But the word *tanoana* does not literally mean "life-force." Kruyt himself explains in fact that it means "manikin" (Adriani and Kruyt 1950, I: 409). It is said to derive from *to ana*, little man, and to be described as "a little man the size of the little finger" (Downs 1956: 33).

As a matter of fact, moreover, Kruyt's own admirably particular ethnography contradicts his interpretation of the mystical efficacy of Toradja head-hunting. Although he explicitly ascribes to them the "idea" of gaining life-force, what they themselves are reported as thinking about the matter is very different. Downs, in his critical analysis of the Toradja ethnographic literature (1956), makes the point that the most important reason for head-hunting, as given by the Toradja, is that if they did not do it their ancestors would punish them by making them sick or spoiling their crops (p. 64). They say that the enemy heads constitute food for the spirits (*anitu*) in the village temple; if the spirits were not given heads to "eat" they would eat the villagers instead and punish them with sickness, troubles, and the deaths of their children. They are also reported as saying expressly that if they did not go head-hunting their crops would fail (Adriani and Kruyt 1950, I: 246). But this is not because they think in terms of maintaining a supply of *tanoana*, conceived as life-force. As Downs stresses (1956: 64), the Toradja did not speak of *tanoana* in this connection:

> The *tanoana* of the enemy were taken to weaken them, not to strengthen their attackers. The fact alone that a single head was sufficient for victory indicates that it was a symbolic act and not a matter of gathering a life-giving substance.

Nevertheless, the association between the taking of heads and the acquisition of all forms of fertility and well-being remains widely and convincingly testified to. For instance, the Sebop of the Tinjar river, Sarawak, told Furness that head-hunting brings "blessings, plentiful harvests, and keeps off sickness, and pains" (1902: 59); and the Konyak Naga, of the Indo-Burma border, assured von Fürer-Haimendorf that "when we captured heads, then we had good harvests, then we had many children, and the people were healthy and strong" (1939: 69). Such ethnographical statements can readily be paralleled from all around the world, and there is no doubt concerning the general association between the practice and its supposed consequences. The more fundamental issue, though, is how this connection is to be understood.

Hutton, in a disquisition on a primitive philosophy of life, with special reference to the Naga peoples, is thoroughly in accord with the view adumbrated by Kruyt over thirty years earlier: the Naga head-hunter "seeks to lay in a supply of life-fertilizer for the benefit of his community, bringing it in the heads of his enemies, from which it exudes into the sacred stones of his village, to pass into the cycle of life of its crops, its livestock, and its human population" (1938: 12).[2] This is a conception which has long been current in anthropology, and which we have seen expressed in the account given by Izikowitz of the reasons for head-hunting among neighbors of the Lamet.

The general train of ideas is as follows: Living beings are animated by souls; the activity of the soul is concentrated in the skull, which serves as a container of life-energy; this source of life can be arrogated to the purposes of another party by removing the head from the body and conveying it to another place; the skull then serves, after ritual treatment, as a kind of reservoir of soul-substance, which emanates over human beings, their livestock, and their crops.

The connection thus seen between head-hunting and welfare is the causal relation:

skull→(soul-substance)→fertility, etc.

The question that has occupied me, however, is whether this is a correct conception. I do not at all doubt that there is a causal connection of some nature between the terms; i.e., that the taking of the head procures or sustains the well-being of the possessor. But what seems to me very much a question is whether this causal relation is effected or mediated by a quasi-physical

---

2. A few pages later, Hutton (1938: 15) expressly cites Kruyt on "soul-substance" in Indonesia. Kruyt, in the source there adduced, writes: "Many customs show that the Indonesians consider the human *head* to contain soul-substance. The great object of head-hunting is to possess themselves of their enemy's soul-substance, in order to increase their own" (1914: 233).

"soul-substance" (life-force, etc.) which can be accumulated or diminished and which pervades its surroundings with a beneficial influence.

III

The Kenyah of Borneo offer a paradigm case for the investigation of this problem. They are well known in the ethnographic literature for their former head-hunting, and the associated trappings are still sources of pride and aesthetic pleasure to them. There is the great advantage, also, that their civilization has been minutely described in many regards, by J. M. Elshout, in what is some of the most intensive ethnographic reporting from anywhere in Borneo. For the present purpose his major monograph *De Kěnja-Dajaks uit het Apo-Kajangebied* (1926) provides crucial evidence, both factual and analytical.

Elshout raises the question why the Kenyah practice head-hunting and why especially they take the head back home with them. To this question, he says, no direct answer can be given; but he nevertheless goes on to write (p. 210), without qualification:

> ... Most certainly the Kenyah believes in a strong magical action which goes out from a head which has been taken, and which can thereby act to strengthen the individual.

The head, Elshout continues, produces this effect to an even greater degree than do other "magical objects," such as iron, blood, antique beads, and so on.

Yet when it comes to the evidence for this idea, Elshout concedes (p. 211) that:

> The information to be had from the Kenyah himself, when one asks him why head-hunting played quite so important a part in his past, and on what grounds the worship of heads really rests, is very scarce, and one is surprised to find that he can say practically nothing about it.

What the Kenyah usually say, it appears, is merely that head-hunting is a *lalìh tjèn těpon*, i.e., a prescription ordained by the ancestors, and that in former times if they did not take heads they were punished by being turned to stone. Even the head-hunting ceremony (*mamat*) casts insufficient light on the origin of the practice: "the idea has for the most part been lost."

Elshout adds, however, that in a certain sense the head serves as a hunting-trophy, and perhaps also as a proof by means of which the head-hunter might convince others of the heroic things that he had done. This latter idea, that the head constituted an undeniable evidence of courage, was found by Elshout to be clearly expressed in his day (p. 211). It is not only a matter, though, of convincing other people that an expedition has been successful: more importantly, it seems, the head is a sign that the warriors enjoy the favor of the *bali akang*, spirits of courage. Men in particular are dependent on these spirits, and it is thanks to this spiritual support that they are able to protect themselves, their womenfolk, and their children from possible attacks by other longhouses or tribes (pp. 212, 213). By Kenyah standards, says Elshout, a head-hunting expedition calls for a great deal of courage; and "a successful head-hunting party is thus a proof that the *bali akang* are favorably disposed towards the tribe" (p. 213). According to one statement, the soul (*beruwa*) of the head actually goes to call the *bali akang*;[3] so that, as Elshout infers, the soul of the victim is taken back to the warriors' village in order to serve, in the eyes of the spirits in question, as proof of victory (p. 218).

The outcome, then, would appear to be that the heads are taken, not for their own sake, but as tokens of the favorable disposition of the spirits upon which the well-being of the head-hunters is conceived to depend (p. 216). Elshout goes on to observe that it is "only a small step to symbolize the beneficial influences, which are thought to be consequences for the tribe of this relationship, in the head itself." According to this interpretation, "the head is viewed as the bearer of the beneficial influences which are promised by the favor of the *bali akang*; hence, a powerful action is exerted by the head," which fortifies the sick and the aged (p. 216). The presence in the vil-

---

3. Elshout writes that it goes to call or send for (Dutch *roepen*) the spirits. The Kenyah statement that he quotes has "*tei . . . alla' bali akang*," which I should render more particularly as to "go and . . . bring the spirits of courage."

lage of a freshly-taken head is indeed explicitly said by the Kenyah to "strengthen" (muhat) everybody in it; the head is
swung over old and bent men in order to "bestow fresh
powers" on their souls (beruwa); and it can "purify" (muhê) the
entire village from all harmful influences.

Elshout sums up that the practice of head-hunting finds its
explanation in "the strong purifying magical action which emanates from the head, and partly also from the victory that has
been gained over the foreign beruwa [soul]." The man who sur-
vives the hazards of a head-hunting expedition demonstrates
thereby that he possesses a strong soul, and "this naturally
benefits the tribe, since it is seen as evidence that the spirits of
courage, the bali akang, are favorably disposed towards the village" (p. 223).[4]

IV

The evidence from the Kenyah thus constitutes a
typical ethnographical instance of head-hunting and its beneficial consequences; but it is typical also in the cast that is given
to the indigenous ideas and practices by the ethnographer's
interpretation.

The Kenyah themselves can say "practically nothing" in explanation of the efficacy of head-hunting. The explanatory
framework in which the facts are presented is largely that of
Elshout himself. About all we know for certain, to go by the
occasional translations of what the Kenyah do say, is that the
taking of heads "strengthens" and "purifies" the possessors of
them. We are supplied with no linguistic evidence that the Kenyah think of the custom, even though Elshout asserts that
"most certainly" they believe so, in terms of any "magical
action" or "beneficial influence" which produces these effects.
Even the practice of swinging a freshly-taken head over an old
man does not demonstrate so much: we are told in that connec-

---

4. There is much else in Elshout's monograph on head-hunting, particularly on the head-hunters' ceremony (mamat), which could well be further expounded; but my present intention is to supply no more than the main features
of the ethnographer's understanding of the custom.

tion that to do so strengthens the soul of the subject, but this does not imply that the head (or the action) does so by virtue of a beneficial operation of a "magical" kind. The well-being of the community is clearly conceived as being causally linked to the practice of head-hunting, but the advantageous consequences are not, so far as the evidence goes, directly procured by the mere taking or possession of the head. In particular, it is not the soul of the victim, or some other spiritual factor pertaining to this, which is the causal agent. The protection and other benefits gained are instead expressly in the gift of the spirits concerned, and the only part that is ascribed to the soul associated with the head is that it acts as a messenger to these spirits.

Although the Kenyah case looks a standard one, therefore, and is indeed especially well described, it furnishes no evidential support for the notion that head-hunting secures prosperity by the acquisition of anything like soul-substance. Rather, it emphasizes the crucial importance of the interpretation that the ethnographer gives to the evidence. Elshout says, for instance, that it is only a small step to "symbolize" the beneficial influences by means of the head itself, and this may well be considered a moderate and plausible construction to place upon the facts that he reports. But he then immediately reverts nevertheless to a substantialist interpretation when he goes on to write that the head is regarded as the "bearer" (Dutch *drager*) of the influences in question, as though in some quasi-physical way it actually conveyed (or even contained) them. And it is particularly revealing, furthermore, that he at once concludes that "hence" (*daarom*) a powerful action is exerted by, or emanates from, the head.

It was with such skeptical reflections in my own head, though I suppose they were not put so clearly, that some years ago I approached the Kenyah of the upper Baram river in Sarawak.[5] There were still at that time old men who claimed to have

5. My researches in Borneo in 1951–52, with a brief return visit in 1955, were carried out with a senior studentship awarded by H. M. Treasury Committee (the "Scarbrough" Committee) for Studentships in Foreign Languages and Cultures, under the auspices of the University of Oxford. Further research in 1958 was made possible by a research fellowship awarded under the Cultural Relations Program of SEATO. For the latter subvention I am greatly indebted to the kind support of the Rt. Hon. Malcolm MacDonald, O.M., P.C.,

taken part in head-hunting expeditions; and I remember viv-idly the gruesome mimicry with which one of them enthusias-tically enacted the swivelling eyes, contorted features, and suddenly twisting collapse of a victim whose head had been lopped from his shoulders as he stood. Rumors were current also that there had been some recrudescence of the practice during the Japanese occupation, and it was said that some of the skulls hanging in certain longhouse verandas were those of Japanese soldiers and of Allied fliers. In spite of the long-standing ukase of the Brooke administration and then the co-lonial government of the day, coupled with the dissuasion ex-erted by the Christian missions, head-hunting was a subject that many Kenyah were very ready to talk about. The tradition was a source of bloody-minded pride to some of them, and I have listened to Kenyah from Long San threaten Penan in my company, in an evil jocularity, with the toll they would take of their heads if only they were not forcibly prevented by outsid-ers. The Penan, for their part, and particularly those near the border with Indonesian Borneo, were genuinely apprehensive about the possibility that head-hunters might really come over from the Apo Kayan and attack them. In the past they were tempting prey to the settled tribes, especially the Kenyah and the Iban; and when I lived with them they often speculated fearfully about strange tracks in the forest and about uniden-tified noises at night near their camps. As late as 1952, the Penan Seliu could still be so terrified by the conviction that head-hunters were after them that they abandoned their household goods and the forest products they had accumu-lated for trade, and for five days fled in panic to the Silat river and thence to the protection of Kenyah overlords at Long Muh, on the Baram.

This was the setting in which, as occasion offered, I put my questions about the causal efficacy of head-hunting to mem-

---

sometime Commissioner-General for the United Kingdom in Southeast Asia.

My first intention when I went to the interior of Borneo, inspired by reading Elshout in particular, was to study the mystical ideology of the Kenyah. I gave up that plan, and embarked instead on what was originally the subsidiary proj-ect of working with the Penan, when I discovered on the spot the large extent of the conversions made by the Christian missions and the consequent desue-tude of Kenyah ritual.

bers of various Kenyah longhouses in the upper reaches of the Baram. The most rewarding place for this purpose was Long Muh. The Kenyah there were adherents of the revivalist Bungan cult, and thus the last practitioners in the area of some form of traditional Kenyah ritual: the remains of sacrificed chickens, impaled, putrefied in front of the longhouse, and in 1952 there still stood a huge and most impressive *keramen* (cf. photographs in Elshout 1926, facing p. 288) as a monument to the last *mamat* ceremony. My inquiries were made in the course of brief visits on the way to or from government-supervised trading-meetings with the Penan, and had always to be subsidiary to my questions about relationships between Kenyah and the forest nomads. The language in which we conversed was usually Malay, in which the Kenyah were fluent after their own fashion (Needham 1958), and later on I was able to supplement questions in this language with glosses in Penan, which is closely cognate to the congeries of Kenyah dialects.

The answers that the Kenyah gave to my questions about the reasons for head-hunting were naturally more or less well informed, and they varied in detail, but the general tenor of the ideas they expressed was quite uniform. The standard response, as Elshout had found over the mountains at Long Nawang, was simply that it was a custom decreed by their ancestors (*tepun*). A further explication was that if previously they did not take heads, the spirits (*bali*)[6] would be angry with them and would punish them with sickness and death. But the crucial and positive idea, repeated to me at different places and on many occasions, was that head-hunting (*ngayau*) brought fertility and a general prosperity in the forms of gongs, pigs, wild game, abundant crops, and all else that the Kenyah desired. In particular the emphasis was laid on the conviction that, in the words of the renowned old chief Mapék Arang as he sat near the skull-rack[7]

6. I did not explicitly note down at the time the class of spirits concerned, and can now only assume that they were the *bali akang*. This is certainly what the Penan said in expounding the Kenyah practice; only as they themselves had never been head-hunters, the *baléi akang* (as they say it) were, significantly enough, a class of purely maleficent spirits.

7. This contained, incidentally, only six skulls, one of which was that of an orang-utan. The Long Muh Kenyah used to have many more, I was told, but Mapék Arang said they had been obliged to throw them away when they fled from their former settlement at the mouth of the Lua river, in the Peliran, be-

in the gallery at Long Muh: "When we took heads we had many children." A number of old men deplored to me the enforced abandonment of the custom, on the ground especially that after they had been obliged to give it up the numbers of their people had diminished.[8] This was the constant theme: head-hunting led to increase and well-being; and its discontinuance had been followed by infertility and misfortune.

The causal connection between head-hunting and prosperity was thus made perfectly plain to me, and there could be no doubt that in the minds of the Kenyah the practice had led directly to what they conceived and desired as its beneficial consequences. But what could never be made plain, by any means of analogy or linguistic equivalents or in any way whatever, was just how the cause produced the effect.

Every attempt that I made to elicit an explanation on this score ended in complete failure: I could not satisfy myself that I had a systematic idea of the causal nexus, and the Kenyah seemed not to grasp what I was after. Admitted, the fleeting circumstances were not propitious to the understanding of a mystical ideology, and I was both materially ignorant and linguistically hampered, but I think that if the Kenyah with whom I spoke had themselves had any clear ideas on the issue then I should in some way have been able to elicit them. Gradually, in the event, I became quite firmly convinced that there were no such ideas, in the form of collective representations, to be discovered. The benefits of head-hunting were not procured, at any rate, by the souls (*beruwen*) of the victims, whether as individual agents or as a more general source of mystical activity. In particular, as I grew assured, there was no notion that the heads were repositories or bearers of anything resembling "soul-substance" or "life-energy" or any comparable sort of quantifiable mystical factor; nor was there ever the slightest indication that the Kenyah thought of the skulls as emitting any kind of force over the people and things in their environment.

---

cause of attack by the Iban. (They also had to abandon other substantial and precious possessions, such as the longhouse slit-drum and even their gongs.)

8. As to the demographic facts, I suppose that the incursion of exotic diseases carried by Europeans and other travelers from the lowlands, as well as probably the infertility consequent upon venereal disease, could have led to a real decline in population during the period at issue.

An evident conclusion, therefore, is that when the Kenyah could tell Elshout, to his admitted surprise, practically nothing about the conceptual foundations of head-hunting, this was because there was really nothing further to tell. The taking of heads secured well-being, and that was that.

V

The Kenyah case thus accretes to that of the Toradja, which was so influential in the formation of the idea of soul-substance, in demonstrating that for head-hunters there need in fact be no intermediary term or factor in the causal connection postulated between taking heads and gaining prosperity.

We may add, also, that there is as little evidence that the Naga peoples entertained such an idea; and a wider purview of the ethnographical literature on head-hunting peoples would only confirm, I think, that the acquisition of a quantifiable form or medium of mystical energy is not a characteristic feature of the ideology behind the practice. It is entirely possible, of course, that in some head-hunting tradition or other there could nevertheless be found a concept corresponding to soul-substance. It might be very like the Lamet idea of *klpu*, for instance, only connected with the taking of human heads. But that would be a matter of evidence which so far as I know has not yet been established. Moreover, even if some such case could be demonstrated, it would not invalidate the present contention that the causality of head-hunting does not require an intermediate variable.[9]

---

9. Dr. P. G. Rivière has since (September 1975) directed me to the Jívaro case, as reported by Michael J. Harner in *The Jívaro: People of the Sacred Waterfalls,* which comes exceedingly close to the conventional anthropological paradigm of head-hunting.

In order to be secure, and to attain true manhood, a Jívaro has to acquire an *arutam wakani,* vision soul: this endows him with a "power" called *kakarma* which gives him a forceful personality and preserves him from death by physical violence or sorcery (p. 139). When such a man is nevertheless killed by either of these means, including head-hunting, he forms a *muisak,* "avenging soul," which attempts to kill the murderer or one of his close relatives (pp. 143–

It is not simply the patent significance of ethnographic facts, or else the lack of certain explicit data, that has led to this realization. A crucial point of method is the reconstruction of the premises adopted by the ethnographers when they put their questions, for the questions may in part have determined the answers or the construction that has been put upon these. When Izikowitz could get no information from the Lamet in support of his supposition that the skull might in some way be regarded by them as "a center of power," he thought he might have drawn a blank because he placed his questions from the wrong angle. "Asking questions is often like adjusting a radio to short wave. One should almost know the answer beforehand, and get the 'station' right away. If the question is a little askew, no answer is obtained" (1951: 334).

---

44). A head-hunter cuts off the head of the victim and by various techniques, culminating in shrinking it into the famous *tsantsa*, seeks to contain the *muisak*: also, the facial skin is rubbed with charcoal so that the avenging soul cannot see out (p. 145), and sometimes a large hard seed is placed under each eyelid, filling any slight aperture that may remain (pp. 188–89). During three feasts which subsequently celebrate the killing, the participants are concerned not only thus to contain the power of the *muisak* but also to utilize it (p. 146):

> . . . The *muisak* emits power, but it is believed that the *muisak's* power is directly transmissible to other persons. The man who took the head holds the *tsantsa* [shrunken head] aloft in the ritual dance, while two female relatives whom he wants to benefit . . . hold on to him. In this manner the power of the *muisak* is believed to be transmitted to the women. . . . This power, transmitted from the *muisak* to the women through the "filtering" mechanism of the head-taker, is believed to make it possible for them to work harder and to be more successful in crop production and in the raising of domestic animals (p. 147).

It is not expressly stated that the power derived through the head is the "impersonal power, *kakarma*, which resembles, but is not precisely identical to, the Oceanian *mana*" (p. 152), but the Jívaro ideology does seem to provide, in Ecuador, an instance of the kind of mystical force that I have called into question in Southeast Asia. It would be most valuable, then, to be provided with a critical gloss on the concept that the ethnographer translates as "power," together with further details on the connection between the *arutam* soul and the *muisak*, and, in particular, to have explained the precise way in which the Jívaro think the power of the *muisak* is transmitted via the head-hunter to the women. For the present, I note only that the mystical force in question does not emanate directly into the crops and animals, but that it makes the women affected able to work harder.

What questions in particular were put to the Toradja, the Kenyah, and the Naga by the ethnographers cannot precisely be determined, but it is a clear inference from their reports that they were seeking an effective variable between cause and effect. My own questions, certainly, were posed with this scheme of ideas in mind: what I wanted to know from the Kenyah was "how" the taking of heads (over and above the simple fact that they were taken) secured well-being, "what made" the heads increase fertility (over and above the assumed fact that they did), and so on. I was puzzled that the Kenyah could give me no corresponding answers to queries phrased in these terms, and that they were satisfied to assert in effect that the practice just did have beneficial consequences; but it was this puzzlement which led me to what I came to regard as the right construction to put on the case.

The common premise on the part of anthropological commentators on head-hunting has been that in the clearly causal connection between taking heads and securing prosperity there must subsist or intervene some forceful medium which brings about the effect. Now indigenous statements that head-hunting procured certain desirable consequences had, as they stood, the simple form:

$$a \rightarrow b$$

But these evidences of how the practitioners conceived the causal nexus were interpreted by the inquirers in the augmented form:

$$a \rightarrow (x) \rightarrow b$$

It thus appeared, consequently, that there was a problem: namely, to isolate the mysterious factor x. The solution was found in the postulation of a medium of mystical energy such as "soul-substance," "life-force," "life-fertilizer," etc.

This formulation, however, was in the first place an offense against Occam's Razor. If the Toradja or the Kenyah or the Naga say something equivalent to $a \rightarrow b$, and nothing more, then there are no logical grounds to interpolate a third term, and to do so is indeed to multiply the entities beyond necessity. Nor, on the evidence supplied, is there any ideological necessity to conceive an intermediate variable. We can already understand perfectly well that for the Toradja and the Kenyah at least it is certain spirits, with proper names and recognized powers, which actually provide the desired consequences of the practice of head-hunting.[10] Even when we have no such indigenous explanation, as in probably the majority of instances reported, there is still no ground to assume that the ideology of head-hunting, as represented by the practitioners themselves, is in some central respect logically defective or incomplete as a causal account.

It might be maintained, rather, that the immediate causal efficacy attributed to the taking of heads is indeed characteristic of a certain mode of thought. Thus Lévy-Bruhl writes in his *Carnets*: "The question of how is not posed for the primitive mentality when what is at issue is a cause belonging to the world of mystical experience" (1949: 26). But this suggestion does not quite fit the present case, for the causality of head-hunting is very clearly stated by the Kenyah, at any rate: for them, the taking of heads is the direct cause of those advantages which they regard as its effects. Furthermore, the causality is confirmed for them by a reasoning, based on concomitant variation, which is entirely "scientific": the cessation of head-hunting is the direct cause of the loss or diminution of the advantages. On this score, we could hardly ask for better proof that in this paradigmatically "mystical" sphere of ideas they are indeed concerned with the "modality" of the events in question. It is just that to their minds the causal nexus does not entail the postulation of an intermediary factor. The cause is unmediated in its effects.

This conclusion does not mean that the Kenyah employ only a single undifferentiated concept of cause, as exhibited in their ideas about head-hunting. We do not know, except as in the present case by the investigation of particular instances and institutions, what causal connections the Kenyah conceive to exist among whatever they discriminate as phenomena. And even a well substantiated explication of their conception of the

---

10. It is interesting that the Lamet, in those practices which do involve skulls, also conceive the effective agencies to be spirits: the skulls of game are hung in the men's house, and the spirit of the forest is enticed into them (Izikowitz 1951: 196–97, 333); the skulls of sacrificial buffaloes are displayed, and the ancestral spirits come and live in them (p. 334). The skulls secure the favor of the spirits, on which men's fortunes depend; and it is the action of the spirits, with their human-like characters—not any mystical influence such as *klpu*—that is the force invoked.

causality of head-hunting would not entail that they possessed no further concepts of the relations between cause and effect in various other circumstances.

Lévy-Bruhl, indeed, has urged that our own ideas about causality are so specific to a western theory of knowledge, as elaborated from Plato and Aristotle down to Hume and Kant, that when analyzing alien traditions of thought (such as that of the Kenyah) we should find terms other than those that are usual in our epistemology. There is no way of doing this, he contends, other than by attempting a description and analysis of the facts, taking these in a way free of all contact with psychological and philosophical speculation on the question, and independently also of what is implied by the associated terminology. "We still have to use the word 'cause.' But let us avoid supposing that the primitive mentality uses it with the atmosphere that it has for us, and let us try to find out precisely what goes on in them (the primitives) when they use their corresponding word" (1949: 35).

How far we are from that stage of investigation is shown by the fact that we do not yet know whether there is a Kenyah word corresponding to "cause."[11] Nor do we possess a range of causal statements in Kenyah which, taken together with their contexts of use, could be material for a more abstract conceptual analysis. Although the notion of cause has been central, since the nineteenth century, to anthropological disquisitions on primitive thought (cf. Needham 1972, chap. 10), the modes of causality as actually conceived in alien traditions have yet to be determined; and even though the institution of head-hunting has for decades served as a dramatic and notorious illustration of a primitive belief about cause and effect, the causal nexus has been constantly misrepresented by anthropological commentators.

VI

The interpretation of head-hunting thus appears more a failure of European reason than a brutal misconception

---

11. To judge by what I know of the related Penan language, I should think there was no such verbal concept, though there are surely (as in Penan) various means of expressing a causal inference or connection.

on the part of those who take heads in order to gain life.

In spite of the lack of direct evidence for the concept of soul-substance, and in the face even of positive evidence that the benefits of head-hunting were thought to accrue from quite another source, the anthropologists have persisted in interpolating a fictitious entity between the cause and the effects. The question then is why they did so, for it is not only a lapse from the principle of Occam's Razor that is responsible, but it seems to me that a conceptual error is also at issue, and the roots of this have to be uncovered. For "in order to convince someone of the truth, it is not enough to state it: we must find the *path* from error to truth" (Wittgenstein 1967b: 234).

A patent attraction in the postulation of a mysterious factor *x* (soul-substance, etc.), and what may well have provided a pragmatic conviction of its appositeness, is that it makes coherent a number of disparate aspects of head-hunting which recur as standard reports in the ethnographic literature:

1.   An effective agent appears to be conveyed, together with the severed head, from one place to another; i.e., an entity with properties distinct from those of the grisly trophy itself is in question.

2.   This entity is immaterial, but animating; therefore like a soul.

3.   It is not however like a personal soul, with volition and other such individual characteristics, but has a uniform nature.

4.   It is quantifiable, and there can be more or less of it; heads are taken not just once but repeatedly, as though a "supply" (Hutton) had to be kept up.

5.   It is concentrated in the skull, as though this served as a container; under this aspect it resembles a fluid.

6.   It requires, together with the head, a fixed and appropriate place (skull-rack, head-tree) from which it exerts its effect; the skull can be taken elsewhere on occasion, but is carefully returned to its special location; this place is central and in other regards important; it not only marks the focus of the community but makes the skull appear a kind of generator of an influence which pervades the area around it.

7.   It acts invisibly and at a distance on its environment, as though it emitted a force.

8.   When the head is carried over growing rice, the crop flourishes, and when it is waved above the head of an old man his strength is restored; so it seems to have a quasi-material aspect, as though it settled on or enveloped (like mist) the things on which it acts.

All of these aspects, as well as other particulars which could be adduced from individual societies, can be brought together in a unitary explanation by means of the assumption that the effective agent in the consequences of head-hunting is a substantial vehicle of "life-force." It is not surprising, therefore, that this received idea in anthropology should have become so readily established, and especially when its proponents included authorities such as Kruyt and Hutton who had lived for years among head-hunters. Nor is it in the least difficult to appreciate on what grounds Izikowitz assimilated the Lamet concept of *klpu* to the idea of soul-substance, for the *klpu* of rice (though not that of man) answers exactly to the properties attributed to this mystical entity.

But the particular terms in which the qualities of *klpu* are described have in themselves a crucial importance, in that they correspond to certain more general conceptual predispositions—and it is these, I shall argue, which are ultimately responsible for the postulation of a fictitious and unnecessary factor. Here we approach what I regard as the most fundamental part of the present investigation, for the concoction of the idea of soul-substance is not sufficiently explained merely by the fact that it proved convenient in analysis. What we need to understand, if we are to derive the fullest instruction from this case, is why just this particular idea, framed in certain characteristic terms, came to be formulated for the purpose.

The answer can be discerned, I think, in the descriptive images and abstractions to which the ethnographers typically resort in their attempts to convey what they conceive as the causal agent at work. Kruyt introduced the idea of a life-"force," first thought of as a "fluid" and then as a "substance" (1906: 2); Elshout referred to magical "energy" and to a magical "balance" which was affected by the skulls (1926: 217, n. 1);

Hutton wrote that the life-fertilizer in the heads "exudes" into their surroundings (1938: 12); and Izikowitz writes of the *klpu* that, rather than resembling a specter of the dead, it is more "a sort of fluid, like electricity" (1941: 9), and that quantities of *klpu* "ray out" into the fields (p. 23).

Now all of these terms belong to a scientific idiom derived from physics: electromagnetism, hydraulics, mechanics. They were adduced by the ethnographers because certain effects had to be causally accounted for; and if the latter were caused by the taking of heads, then these must have produced the effects by means of some "force" or "energy." This is the way physical effects are accounted for, and it was in accordance with this scientific model of explanation that (to judge by the terms typically employed) the ethnographers conceived what they thought a fit interpretation of the causality of head-hunting.

The scientific idiom in question, and the scheme of theoretical ideas that it expresses, pervades the daily experience of technological civilizations. You press a switch and a light goes on: this is because an electric current flows between, through the physical medium of a wire. You strike with a hammer and a nail penetrates a board: this is because a kinetic force is imparted from the tool to the head of the nail. You light a fire and heat is produced in things nearby: this is because radiant energy is emitted by the combustion of the material. A wireless receiver reproduces sounds: this is because radio waves are propagated into space by a transmitter. You approach an automatic door and it opens: this is because your body has interrupted a beam to a photoelectric cell. A navigation guide shines in the dark: this is because it sends out light waves. Examples of this kind are familiar and innumerable, and in each instance a given agent is said to cause a specific effect by means of some form of force or energy. This type of causal explanation has a long history, but achieved its most influential expression in the great scientific revolution of the nineteenth century. It is this triumph of physical science which is responsible for the "tone of thought" (Waismann 1968: 65) that has provided the explanatory model, and the associated idiom, to which anthropologists have resorted in their attempts to explain the ideology of head-hunting.

This is a factor in the history of ideas which can already be recognized as having had an influence of a profound kind on sociological theory. Some years ago I wrote: "It is intriguing to conjecture the effect of nineteenth-century physics on the development of such notions as 'cause' and 'force' in Durkheim's thought, and which led Mauss to look for a 'force' in a gift which compelled its return" (in Durkheim and Mauss 1963: xxv, n. 1). What I had in mind was in the first place Durkheim's constant reliance on physical, and particularly electrical, analogies in making his sociological arguments.[12] Then there was the stress that, together with Mauss, he laid on a presumed causal connection between social forms and symbolic classifications (1903), such that a sociological explanation of symbolic categories was thought to consist in the establishment of systematic connections between these and the modes of social grouping by which they had been determined. Another example from the period is van Gennep's elucidation of the marginal state or point (*marge*) in human activities and in other kinds of events. He begins, revealingly enough, with an illustration taken specifically from mechanics, in which it is necessary that two movements in opposite directions shall be separated by a point of inertia. This point can be virtually eliminated by a circular motion, but the case is not the same in biological or social activities, "for these deteriorate and need to be regenerated at more or less close intervals" (1909: 260). It is to this necessity, van Gennep concludes, that rites of passage correspond; and he frames this necessity in terms of mechanics and what he seems to conceive as the replenishment of energy.[13] Later, Mauss provides a typical instance of the idiom at issue when he opens his essay on the gift with the question, which he describes as the problem to which he specially applies himself: "What force is there in the thing given which makes the recipient return it?" (1925: 33).

In one such example after another, and in a train of argument extending into more recent discussions about primitive

---

12. The point has lately been taken up by Lukes in his observations on the figurative language used by Durkheim in characterizing social phenomena (Lukes 1973: 35–36), especially in such phrases as "collective forces," "social currents," etc.

13. On this score, the English edition (van Gennep 1960) goes beyond what is literally justified. "Car celles-ci s'usent . . ." is rendered as: "Their energy becomes exhausted . . ." (p. 181). But this very license on the part of the translators goes to confirm my case.

thought and western rationality, the idiom of scientific method has in this way tended to bias the interpretation of alien ideologies. With the progress of science, the idiom has acquired new, post-Newtonian terms, particularly those elaborated in quantum mechanics; but this development is likely only to have confirmed the tendency to conceive conceptual problems in terms borrowed from the most dominant and successful of the sciences. The theories behind the vocabulary of forces have become more subtle and obscure, but the idiom has not for this reason become the less attractive to social anthropologists seeking explanatory models.

In the note cited above, I merely suggested that it would be intriguing to conjecture the effect of nineteenth-century physics on sociological thought. I did not, in an incidental observation, try to trace the actual influence of that science on the *Année sociologique* school, and this is a task which still remains to be carried out. But the present case has, I suggest, now supplied a real example of what I was talking about. It was the idiom of physics, I have argued, which gave ethnographical inquiries into the causality of head-hunting such forms as "What makes head-hunting secure prosperity?" or "What is it in the head that gives life?" and so on, in the same expectation that the causal activity (taking the head) procured certain effects by means of some kind of force or energy. Not that such questions are in principle unjustified or in practice useless: they could correspond to the way head-hunters do think, and they would thus elicit what was crucial in their ideology. But it is worse than useless to keep pressing the inquiry in such terms when the answers do not justify them; and it leads only to confusion and miscomprehension then to concoct an unevidenced notion which corresponds not to the answers received but to the questions put.

In the present case, it is not merely that the questions are a little askew, so that they fail to correspond to ideas which the head-hunters actually hold and otherwise might express. The very form of the questions is the product of certain theoretical preconceptions which are not part of the ideology of head-hunting. To the practitioners, therefore, the questions are not intelligible and cannot be answered. The ethnographers, failing to elicit an idea (the mysterious factor x) answering to their own expectations, postulate an abstraction (a mystical kind of force) that satisfies what for them is a necessary condition of the relation between cause and effect. But the reality of the issue is that underneath the mutual bafflement there is a confrontation between two quite different conceptions of causality.

VII

Finally, there is one further consideration that I wish to mention, not only as a matter of good method but also because it seems to me to have a more profound importance than the particulars of the problem that we have been investigating. It has to do with the deliberate improvement of our own thought.

In the present essay I have isolated a defective theoretical conception, and I have gone on to trace the origin of that defect. If the argument is accepted, then a definite advance in understanding has been secured: we can now more easily recognize that in symbolic thought a given cause can produce an effect directly, without the mediation of anything resembling a force;[14] and we can also be more on our guard against any explanatory bias that may be imparted by the scientific idiom in question. But a deeper concern yet should now occupy us. If such admirable ethnographers as Kruyt and his successors labored so long under a misapprehension about causality, what about ourselves?

It is an unavoidable inference that we too must be thinking about social facts in comparably invalidating ways, and that in our case also the influences responsible can in principle (if not now) be identified. We therefore need, in our turn, to appreciate the "tone of thought" which characterizes our speculations, and to take into account its constraints. Obviously, this radical kind of critique is the hardest of intellectual undertakings, for

---

14. The idea presents itself that this may prove a significant point in a more general study of ritual, i.e., symbolic action, especially such practices as blessing, cursing, purifying, etc. But before this possibility can be estimated we shall need to know a great deal more about alien ideas of causality in such everyday pragmatic contexts as carving, hammering, grinding, and so on.

we cannot by deliberation alone detach ourselves sufficiently from those tacit premises which themselves frame or constitute our thoughts—but we have to try. Let me make just a few brief observations in that direction.

Perhaps an apt beginning is made by the very topic of mechanical models. Wittgenstein, when analyzing a certain puzzle in the concept of willing, stresses the effect of a misleading analogy of the kind: a causal nexus seems to be established by a mechanism connecting two parts of a machine, such that the connection may be broken if the mechanism is disturbed. And he then adds the parenthetical comment: "We think only of the disturbances to which a machine is normally subject; not, say, of cog-wheels suddenly going soft, or passing through one another, and so on" (1953, sec. 613). In the present perspective, the point that this striking image makes is not only that a mechanical model leads us to think of fixed points of connection and regular modes of operation in whatever we try to understand systematically, but also that the idiom encourages a *rigidity* of outlook which is quite inappropriate to the subtle interplay of ideas.[15]

This tendency is aggravated by a conceptual predisposition concerning what constitutes a class, and hence the definition of those variables that we try to correlate or otherwise account for. This analytical hazard is found in its most deleterious form in the assumption that a conceptual class is a number of individuals known by a common name by virtue of some respect in which they are alike. According to this standard definition, when we describe a class we specify what is common to all members of it. But Wittgenstein has abundantly proved that this is not necessarily the way that conceptual classes are actually formed; and Vygotsky has shown experimentally that this mode of definition is not always the pattern on which concepts are in fact formed or apprehended (cf. Needham 1972, chap. 7). These conclusions can have critical consequences for employment of the classificatory concepts upon which social anthro-

---

15. On the score of expository idiom alone, it may be wondered exactly how appropriate to Jívaro thought are the terms to which Harner resorts (1973: 147) when he states that the *muisak* (cf. above, n. 9) "emits" a "power" and that this can be "transmitted" through a "filtering mechanism."

pologists have been accustomed to rely (Needham 1975). I have tried to demonstrate this by reference to the standard concepts of kinship studies (Needham 1974a, chap. 1). Wittgenstein has ascribed the defective definition of a class, in the first place, to a craving for generality which inclines us to assume that there must be something in common to all the entities that we commonly subsume under a single term; and secondly, to our preoccupation with the method of science. These related predispositions tend to frame premises which, though undeclared, can very much mislead us in the analysis of social facts. Both of them have been at work, I should contend, in the explanation of head-hunting.

Underlying the rigidity of thought, and the generalizing influence of the classificatory instruments of thought, there is a body of assumptions about what constitutes human nature, in particular about the essential capacities of the thinking subject. Social anthropology has in the main been founded on the presumption that these were already known, and that the rational and psychological vocabularies of European languages were well adapted to the comprehension and analysis of alien modes of experience. In an inquiry into received ideas about the supposed capacity for belief (Needham 1972), I have tried to demonstrate that we have no such certainty; and that a distinctive task of social anthropology should be to determine whether there are to be found any absolute features of thought and action—in the form of natural resemblances throughout mankind—that are indispensable to an objective conception of humanity. In this field practically everything remains to be done, and until we have elaborated and justified a set of appropriate categories it will scarcely be feasible to interpret the causality of head-hunting, or any other ideological institution, with either plausibility or exactitude.

At present we are at the prior stage of the "undermining of categories over the whole field of thought" (Waismann 1968: 21). This task has been embarked on with respect to the analytical concepts of kinship studies and the psychological concept of belief, and has now been repeated in the present disintegration of the idea of soul-substance and the model of causality with which it is connected. So these various examples alert us to certain improper suasions by which our thoughts may be directed; and they show us how to proceed in a sustained undermining of some categories and the better establishment of others. But such defective premises are hazards that we can already discern, and this means that we have not yet reached that level of abstraction at which a "tone of thought," such as organizes and directs the ideas of an age (Waismann 1968: 65), will exert its more insidious influence. Moreover, the grounds of these defects can be traced to identifiable determinants: classical physics, for example, or a parochial philosophy unchecked by a confrontation with other linguistic traditions and exotic social facts. But if (as I assume) we are at present misguided by subliminal attitudes of mind which "hold thought in a firm position" (Waismann 1968: 216), it is to be imagined rather that these constraints will be far more subtle, profound, and pervasive.

In such regards the outcome of the present critique cannot be at all conclusive; but it may show that skeptical vigilance and the historical tracing of particular preconceptions can prove the means, in Waismann's liberating metaphor (1968: 217), of uncramping our minds and "regaining intellectual freedom."

We predicate of the thing what lies in the method of representing it. Impressed by the possibility of a comparison, we think we are perceiving a state of affairs of the highest generality.

Wittgenstein

5

Reversals

I

In the comparative analysis of social facts, certain more or less formal relational concepts have proved to have a special efficacy in various fields of investigation.

Thus in the study of prescriptive systems a great deal of clarity has been achieved by resort to the relations of symmetry, asymmetry, and transitivity; and in the field of symbolic classification, similar benefits have accrued from reliance upon the notions of opposition, analogy, and homology. The simply pragmatic consequences of the employment of such relational ideas have indeed been quite impressive in the task of rendering certain classes of social facts more readily intelligible. But there remains to be undertaken the more radical task of discovering on what grounds our explications are resting when we resort to such ideas. They are not all self-evidently ultimate terms in the sense that they are logically simple or that they are resistant, at least, to the analysis of their own grounds, constitution, or validity. These relations, and the justification of their qualified utility, are just as much proper objects of comparativism as are the social phenomena which they order.

Some relations have a clear formal definition, and this provides a ground against which social facts can be assessed. An obvious candidate is symmetry: $aRb \sqsupset bRa$. This provides a simple and steady gauge against which to analyze a symmetric prescriptive classification; also to appreciate any contingent nonsymmetrical or asymmetrical features in certain sectors. Similarly, transitivity can be absolutely defined in formal terms: $(aRb) \cdot (bRc) \sqsupset aRc$. This provides a criterion for such matters as estimating the concatenations of asymmetric alliance, the limitations imposed by exogamous moieties in prescriptive systems, or the operation of a social hierarchy.

Other relations are less clear, or less certain, and analyses carried out in terms of them are made correspondingly unreliable (Needham 1980a, chap. 2). The prime candidate in this class is the notion of opposition, despite its pragmatic value in the explication of social facts. By Aristotelian criteria there are four modes of opposition, and three of these are directly pertinent to the analysis of symbolic representations. Ogden distinguished no fewer than 25 modes of opposition in ordinary English usage (1932). In addition, and by other criteria, there are numerous further modes that can be distinguished. The practical efficacy of the notion does not provide a formal justification of it; and there is no logical means of determining which, if any, of the different modes can claim to be paradigmatic or the ultimate type of opposition. Analogy, in its turn, still awaits a definitive formal analysis, and the grounds of its practical use in the construction of symbolic classifications have not yet been thoroughly explored. As for homology, this is a derivative (in the field of collective representations, at any rate) of the relations of opposition and analogy; it is more obscure than either of these, and hence is far less reliable as an analytical concept.

It is with no exaggerated expectations, therefore, that we can now turn to the relation that is the subject of the present examination—that of reversal.

II

In order to remind ourselves of the kinds of social facts in question, and to form a preliminary idea of the relation at work, let us begin by turning to Middleton's well-known account of some categories of dual classification among the Lugbara of Uganda.

Before the formation of Lugbara society, the various mythical personages were characterized by features which Middleton calls "inverted." They behaved in ways that were the "opposite" of what is expected of normal socialized persons in Lugbara society today. They were at first ignorant of sexual intercourse; the blood of the first menstruation of the first woman was not human blood; they committed incest; they ate their own children; they did not know of kinship or marriage; they were inhuman in appearance; they could perform miraculous acts; they lived by hunting. "The concept of 'inversion,'" Middleton writes, "represents the presocial period . . . before there was an ordered society, when there was, instead, a world of human disorder or chaos." A similar pattern, he continues, can be seen in spatial terms. Beyond the field of social relations of any family cluster are groups which are believed to be sorcerers and magicians; and beyond them, along and over the horizon, are thought to live incestuous cannibals "who are said to walk on their heads and to commit almost unimaginable abominations" (Middleton 1973: 372–73).

This catalogue of contrasts makes up a fairly typical example of a symbolic recourse which has been reported from all round the world and from every form of civilization. Despite their cultural singularity, the disparities posited by the Lugbara make an entirely familiar impression, as though we were encountering again a unitary class of social facts. This sense of recognition encourages the conclusion that we have understood something, in that we have grasped a principle of organization by which certain collective representations are ordered. But when we consider the instances more closely, and then try to abstract the precise principle at work, the matter quickly becomes confusing.

In the present example, the ethnographer has chosen to class the social facts in question as instances of "inversion." He might alternatively have labeled them, as comparativists often have done in other cases, as instances of "reversal." Let us therefore open the analysis by examining these two words, their constitution and their common uses.

To "invert" means literally (by its Latin derivation) to turn in, to turn outside in, hence to turn the opposite way; transferentially, it means to turn upside-down, to reverse in position, order, or sequence; to turn in an opposite direction. To "reverse" means, by its Latin derivation, to turn back; it has among its meanings the senses of to turn or place upside-down, to invert; to turn the other way, in respect of position or aspect; to transpose, turn inside-out; to convert into something of an opposite character; to employ or perform in a way opposite to the former or usual method; to turn in the opposite direction.

The substantive forms "inversion" and "reversal" are hence largely equivalent. Inversion is defined by reference to reversal; reversal is defined by reference to inversion; and certain of the directional or operational senses of each word are ascribed identically to the other. It is of special interest, moreover, that both are defined by reference to "opposition." They are not, however, more particular forms of opposition; they are vague and various modes of opposition, a relation which itself is yet more variegated. Etymologically, "opposition" derives from the Latin prefix *op-*, from *ob-*, connoting toward, before, facing, against; added to *ponere*, to place, put, or set; giving the combination *opponere*, to set over against. When applied to other than material objects, the words "inversion," "reversal," and "opposition" are thus not descriptive, but figurative; like many expressions for mental operations, they are spatial metaphors.

Considered spatially, "inversion" and "reversal" do not differ markedly from each other, except that inversion is ordinarily pictured vertically and reversal horizontally. Inversion, however, has a range of specific connotations in rhetoric, grammar, music, logic, mathematics, and chemistry, whereas reversal in the main has not; so, in order to avoid any suggestion of a prior commitment to one or another technical employment, I shall speak for the most part only of "reversal." The question then is how apt this figurative term may be to the comparative analysis of social facts.

The first thing to look for is whether the word corresponds to others in the symbolic or metaphorical discourse of other natural languages, especially outside the Indo-European sphere, but this seems not to obtain at all widely. In fact the only correspondence that I can call to mind is to be found in the language of Kédang, in eastern Indonesia. In discussing incest, the ethnographer R. H. Barnes reports two equivalents. Sexual relations between lines which are related as siblings are spoken of as *hunéq-koloq*, "to turn upside down" (as also, it is added, to turn a house-post upside-down). Another expression, more extreme, is *ula-lojo*, which is said to mean (walking with) "feet to the sky" (Barnes 1974: 260). It does not appear whether either of these expressions is extended to the entire class of whatever the Kédang prohibit or abhor as instances of reversal, a kind of movement which "brings disaster" (p. 306), but Barnes is able to propose as a general formula in Kédang metaphysics: "Well-being depends on the irreversibility of relations; but when things go seriously wrong, one must resort to the opposite principle, inversion." For instance, when one is hopelessly lost on an unfamiliar trail, the only way to get out of such a situation is to put one's clothes on upside-down (p. 306). This is a most interesting case, and it shows that in at least one exotic language what we call reversal (inversion) is denoted by spatial metaphors similar to our own; but at the same time it is a correspondence of verbal imagery which appears elsewhere to be uncommon.

So far as comparative evidence goes, therefore, we cannot at present justify the notion of reversal on the empirical ground that it corresponds to a widespread and perhaps natural linguistic discrimination. Nor can a justification be found, expectably enough, in any significance common to particular symbolic reversals. For example, whereas in Kédang a person recovers a way by putting clothes on upside-down, for a Nyoro woman to do so would show that she desired her husband's death, since in her society it is a sign of mourning (Gorju, cited in Needham 1973a: 308). On the other hand, we cannot justify the notion by remarking how readily the Lugbara and Kédang reports can be accommodated by the variety of connotations under the word "reversal" in an English dictionary; for it is precisely this wide linguistic resource which permits the ethnographic data to be classed as reversals. What we need, then, is an independent validation of this means of constituting a class of social facts.

At this point we need to survey more deliberately the range of social usages which are usually reported or accepted as instances of reversal. This procedure does not imply in advance a specific definition of a concrete class, let alone a typology. I shall be presenting a list of characteristic features, without making any claim that one example is more definitive than any other or that the list is in any respect exhaustive. Only brief ethnographic illustrations will be provided; they are merely reminders of the kinds of evidence available.

III

1. Upside-down. The Batak of Sumatra conceive certain spirits, including ancestral ghosts, as "climbing" head-first when they go down steps (Warneck, cited in Needham 1973a: 307); Kaguru witches are thought to walk upside-down, on their hands (Beidelman 1963: 65); the spirit of a Rotinese who has died a "bad death" will wander about upside-down (Fox 1973: 361); certain Toraja spirits have their noses upside-down (Kruyt 1973: 77), and so does the mythical *mawas* creature in Malaya (Needham 1956: 56, n. 2).

2. Inside-out. In ancient Japan, if one wanted to have a delightful dream one went to sleep wearing one's clothes inside-out (Morris 1979: 112); in Kédang, a witch is described as "outside-in" (Barnes 1974: 211).

3. Backward constitution. In the Bada' district of Toraja country, the dead are thought to have the chest at the back of the body, and the calves of the legs in the place of the shinbones; their faces and feet are turned backwards (Kruyt 1973: 77). In an Irish folktale, the Devil, when bidden by a saint to pray, replies that he cannot kneel because his knees are behind him (Hull 1904: 34, n. 1), and other cases of the kind are reported from Brazil and Argentina; in New Caledonia, a snake changes into a being that looks human, but with elbows and knees backwards and his eyes at the back of his head (Gaidoz 1893: 172). In northern India, the spirit of a woman who died in labor has her feet turned backwards, with the heel to the front and the toes to the back (Crooke 1894: 167).

4. Backward action. In the Huichol land of mythical origin, conversations are conducted with the parties standing back to back (Myerhoff 1978: 227). The ancients of classical antiquity, according to Sir Thomas Browne, kindled the funeral pyre "aversely, or turning their face from it" (1658, chap. 4).

5. Right/left. The Toraja reason that as the living normally do everything with the right hand, so the dead do everything with the left; what for the living is right, for the dead is left; and the living employ the left whenever they do anything for or in connection with the dead (Kruyt 1973: 79, 80).

6. Transvestism. In the *naven* ceremony of the Iatmul, marking the first accomplishment of certain ideal forms of behavior, the outstanding feature is that men dress in women's clothes and women as men (Bateson 1936: 12); the *berdache* of North America dresses as a woman, and so does a Chukchee shaman; Kenyah women bear the proud appurtenances and weapons of warriors.

7. Exchange of attributes. Among the eastern Bororo, during ceremonies connected with certain spiritual beings (*aroe*), there are extensive transpositions of statuses between members of the moieties, and by these exchanges "each moiety becomes the other"; during the ceremonies, "the Tugarege are the Exerae and the Exerae, the Tugarege" (Crocker 1977: 141). In the European tradition there are stock examples of the kind: once a year, Roman slaves issued orders in their masters' households; a boy admiral commands a warship; a boy bishop assumes the mitre and crozier, and for a period receives mock-homage; on Christmas Day, officers in the British Army wait on their men at table and with good grace accept what would otherwise be insolence from them.

8. Lexical reversal. The dead of the Toraja speak the same language as the living, only certain tribes maintain that in the speech of the dead the syllables of words are reversed, in that the first and the last syllables are transposed. Thus when the living say *rano*, lake, the dead speak of *nora*; in the mouths of the dead, *madago*, good, is pronounced *godama* (Kruyt 1973: 78).

9. Semantic reversal. In the speech of the Toraja dead, the meanings given to the words are the reverse of those ascribed to them by the living. "Yes" on earth means "no" in the underworld; when the living say "forwards" the dead understand this as "backwards" or as "stand still." This peculiarity of speech is not confined to the dead, but all sorts of spirits can also be recognized by it (Kruyt 1973: 78). Among the Ngaju of southern Borneo, it is thought that the language spoken in the afterlife is the reverse of the language in this: "right" becomes

"left" there, "straight" is "crooked," "sweet" means "bitter," for "stand up" one says "lie down" (Hardeland and Witschi, cited in Needham 1973a: 307). In the Huichol land of mythical origin, "one says yes when he or she means no." During the peyote hunt, the living describe their circumstances in reversed terms: e.g., after a successful day of gathering baskets full of peyote, one pilgrim, while standing in the moonlight, said: "Ah, what a pity that we have caught no peyote. Here we sit, sad, surrounded by baskets of flowers under a cold sun" (Myerhoff 1978: 228–29).

10. Affront to the customary. The mythical predecessors of the Lugbara perpetrated incest and cannibalism. In Nyoro tradition, the first king of the Bito dynasty, a culture hero called Mpuga Rukidi, is characterized by alien origin, illegitimacy, menial status, savage ignorance, adultery, and other scandalous and sinister attributes (Roscoe et al., cited in Needham 1973a: 317–27). In the Huichol land of origin, a comparable detail is that, in greeting, a person proffers a foot instead of a hand (Myerhoff 1978: 227).

11. Difference from the conventional. Among the descriptions of how the dead of the Toraja "do everything precisely the other way round from the living" are reports of actions which are, it appears, simply different. When the living go to draw water, they lay the bamboo containers on their shoulders, but the dead behave "differently"; they carry the bamboos in their hands. When it was still the custom for men to carry spears as they walked, they carried them on the shoulder, and when they climbed up into a house they stuck the lower end of the shaft into the ground; but the dead behave "differently," in that they hold the spear in the hand as they walk, and when they go up into a house they set the spear against one of the house-posts. The living carry loads in a basket on the back; the dead are "different," and hang the burden from the end of a stick over the shoulder (Kruyt 1973: 76–77).

12. Abstention from normal practice. On Roti, in eastern Indonesia, in the event of a normal burial a payment is made to the mother's brother for ritual services, but in the case of a bad death no payment should be made. After a normal death, the body of the deceased is buried in a coffin, but someone who has died a bad death should not be afforded a coffin. The

Rotinese say that "these inversions 'cause' the spirits . . . themselves to be inverted." In this context a difference between normal and abnormal is symbolized by the performance of ritual
as contrasted with "lack of ritual" (Fox 1973: 359–62).

IV

This list assorts what are described as reversals
(or inversions) into twelve kinds, and these kinds give a fairly
comprehensive idea, I think, of the characteristic features of
the relation in question.

It is true that most of the instances have been separated from
other usages with which they are more or less systematically
connected, but my aim so far has been to distinguish kinds of
reversal, as these are separably given in ethnographic reports
(and just as they may be separately learned in the field), not to
trace their connections with other social facts. It is plain also
that the twelve kinds are not to be taken in a concrete sense,
and that the list does not compose an absolute or exhaustive
classification.

Certain kinds of reversal could be further divided; e.g., under right/left, handedness could well be considered separately
from circumambulation, though the two activities imply
equally a resort to lateral values. Conversely, other kinds of reversal could be aggregated into one: e.g., those which have to
do with directionality. By yet other criteria, furthermore, certain usages could be detached from their respective kinds and
be redistributed among other kinds. In addition, it would be
possible to ransack ethnographic literature and to discriminate
kinds of reversal which are not easily accommodable in our list.

For instance, the Ngaju claim to recognize certain premoni-
tory signs shortly before a death, in that the subject undergoes
a change of character either for good or for ill; someone who
has loved his children turns suddenly against them, while he
who was at odds with his changes into a loving parent (Hardeland 1859: 172, s.v. hawä). And finally there is in any case a possible uncertainty about distinguishing particular kinds of reversal: e.g., the Toraja regard themselves as white in comparison with the dead, who are black (Kruyt 1973: 73), and it

is not obvious how these facts ought to be treated. It is said by the ethnographer that the living are white in agreement with the daylight in which they rejoice, whereas the dead are pitch-black in agreement with the night in which they live; but these reports do not justify the inference that black is the "reverse" of white in Toraja ideology, and even if this supposition were conceded it would still be hard to generalize the opposition of colors into a distinct kind of reversal.

With these qualifications, then, the twelve kinds of reversal do seem to hang together. Certainly ethnographers have taken them to do so, and in particular the accounts of the Toraja, the Rotinese, and the Huichol happen to adduce instances of many of the kinds aggregated together, as similar examples of patterns of thought about spiritual existence and the fate of the living after death. In representing such ideas as reversals or inversions, the ethnographers have at least the support of linguistic usage in Dutch, English, and Spanish; and in some cases, also, the peoples under study apparently regard the various kinds of reversal as having a common meaning or as being at any rate significantly interconnected. Without being more critical of the notion, for the time being, let us take it that we have collocated the characteristic features of a class of social facts conventionally defined by reversal. How, with reference to this relation, are they to be explained? There has been quite a variety of attempts to do so, and our next task is to run quickly through some of the more recent. This will have to mean a rather summary treatment, but the conspectus will help us to find some theoretical bearings.

A beginning is supplied by the Christian legend that St. Peter was crucified upside-down. Jonathan Smith asks, in this connection, "What does it mean to be upside-down?" The answer he entertains is that "its most basic sense is to be nonhuman"; man's proper and distinctive posture is to be upright, and "to reverse this . . . is an annihilation of humanity" (Smith 1970: 290). This is a highly particular interpretation, though extremely general in its terms, and although it fits some at least of our upside-down examples it provides no grounds to expect that it will apply more widely. In the same Christian tradition, indeed, it used to be imagined that the Black Mass was performed by a false priest who somehow was upside-down (and whose back, incidentally, was "turned ignominiously" toward the altar); this sacrilege was supposed to be brought about by the Devil as a derision of the sacrament (de Lancre 1613: 122, 457–66), and its significance lay in this intention, not in the enactment of the "nonhuman" or of the annihilation of humanity. In any event, the criterion of humanity is adduced in only the upside-down kind of reversal, and it does not clarify other kinds such as the exchange of attributes practiced by the Bororo.

At a burial in Kédang the corpse is plundered, and Barnes says that this "must be seen as a reversal of normal decorum." He goes on to argue that the funeral consists of a series of ritual acts meant to mark the progressive separation between the spirit and the body, and such, he thinks, "is the explanation for the reversal" (1974: 185). This is ethnographically informative, but it does not explain why it is that reversal, and not some other conceivable means, is employed for the purpose. More-over, it is made plain elsewhere in Barnes's admirable mono-graph that in other contexts reversal is thought to bring disas-ter and is associated with categorical confusion; though also, as we have seen, when things go seriously wrong one must delib-erately resort to inversion in order to set matters right (p. 306). Categorical separation is indeed crucial to these reversals, but it would not be correct to infer that reversal in general procured a separation, or that this consequence was the explanation of the means adopted. Barnes describes intentional reversal as "a logical technique for putting disorder back in order" (p. 306), but this formulation as well, while perhaps persuasive as an explication of Kédang thought, does not explain why reversal in particular is resorted to, and it does not fit most of the kinds of reversal that we have surveyed. Whether reversal of any kind, in the field of social facts, is a logical operation is a sepa-rate matter to which we shall return.

Barbara Myerhoff, in her discussion of ritual reversal in the peyote hunt among the Huichol Indians, is singularly explicit about what the relation may mean. The function of the rever-sals, she suggests, is to transform the mundane into the sacred by disguising the everyday features of the environment, soci-ety, and behavior, and in some Durkheimian sense setting it apart. They are important in that actions, etc., are altered so that the participants are conscious at all times of the extraordi-

nary nature of their undertaking (Myerhoff 1978: 230). The peyote hunters, "as deities," are going back, and they signify this
by doing everything backwards (p. 231). Peyote pilgrims live
for a time in the supernatural, and "the theme of opposition
provides the details that are needed to make the drama credi-
ble and convincing; the metaphor of backwardness makes for a
concretization and amplification of the ineffable."

Myerhoff sums up by writing that "separation, transformation, and concretization . . . are three purposes achieved by the
reversals. . . . There is a fourth, perhaps the most important
and common function of rituals of this nature. That is the capacity of reversals to invoke continuity through emphasis on
opposition" (p. 232). Nevertheless, she continues, there is no
final significance, for "we continue to unravel additional layers
of meaning, to discover more and more functions fulfilled by
reversals in various contexts." Thus "there is no question,"
Myerhoff concludes, "of looking for the true or correct meaning in the use of reversals. We are dealing with a symbolic refer-
ent that has new meaning in every new context, and within a
single context embraces multiple and contradictory meanings
simultaneously" (p. 235).

These excellent observations call for special attention, and
they will attract renewed assessment as we proceed, but for the
present I shall suggest just a few responses. When Myerhoff alludes to the consciousness of Huichol participants, and to the
credibility and conviction in their eyes of the symbolic drama
which they enact, we can only defer to her authority as an ethnographer, and an impressive one at that. Similarly, when she
implies that Huichol time is linear, and that the past is behind,
we may accept that these attributes express the indigenous idiom, so that it may be seen as fitting that the Indians should do
things backwards in order to represent or revert to the past.
However, when she proposes that reversal makes for a concretization of the ineffable, her analysis can be resisted on its own
grounds; for what is in question is not ineffable but is explicitly
expressed—in the very terms of reversal—and we are given no
evidence that this is merely a metaphorical substitute for other
and deeper meanings which the Huichol are frustrated from expressing. As for the factors of separation and transformation,
these are not purposes achieved by reversal: they are the very

conditions of the reversals themselves, and aspects under which they are brought about. Lastly, the function that Myerhoff thinks the most important—namely, the capacity of reversals to invoke continuity through emphasis on opposition—may be apt within Huichol ideology, but it is ineffectually paradoxical when extended to a global range of kinds of reversal which constantly symbolize significant discontinuities.

Peacock has made an interesting suggestion about reversals in general. Drawing a distinction between classificatory and instrumental world-views, he ventures the ideas that the former, which emphasizes the subsuming of symbols within a frame, "nourishes and is nourished by symbols of reversal," whereas the instrumental world-view, which emphasizes the harnessing of means to an end, threatens and is threatened by such symbols (1978: 221–22). It may be rather hard to detect a specific etiology in these figures of speech, but the idiom to which the analyst resorts is an expression of the intricacy of the problem and connects up with other approaches. Thus Sir Thomas Browne subscribed to a similar idiom when he wrote that "contraries, though they destroy one another, are yet the life of one another" (1643, part 2, sec. 4). For the present, perhaps it can at least be entertained that a symbolic classification calls not only for ritual expression but even for a degree of histrionic exaggeration of its main categories, and that an extreme form of such dynamic sustenance may be found in a resort to reversal.

Next, in a general survey of a volume devoted to the study of symbolic inversions, Barbara Babcock regards reversal as "cultural negation." She first defines symbolic inversion as comprising "any act of expressive behavior which inverts, contradicts, abrogates, or in some fashion presents an alternative to commonly held cultural codes, values, and norms" (1978: 14). Then she proposes that "it is through various forms of symbolic inversion that culture frees itself from the limitations of 'thou-shalt-not's,' enriches itself with the subject-matter without which it could not work efficiently, and enables itself to speak about itself" (pp. 20–21). These figurative expressions reflect a long factual treatment, and it would be inappropriate to contest them on the ground that they are figurative, just as it would be to quibble about the hypostasization of culture. The centrally important points in Babcock's case, in my view, are the idea of "negation" and the variety of acts which are admitted as inversions. We shall take up these points in a moment.

Finally, in this quick parade of predecessors, I want to quote a passage on our theme from Hallpike's challenging discourse on the foundations of primitive thought. "When primitives attempt to conceptualize disorder," he writes, "we frequently find that they achieve this by inverting the existing order of things. . . . The de-structuring of human nature leads to notions of inversion and reversal" (1979: 460). It is not necessary, however, to imagine primitives putting themselves to the detached philosophical exertion of trying to form a concept of disorder; and although it is true that people commonly imagine a contrast with a prevailing order by means of inversion, what results from the operation is not disorder: it is a different form of order. Even if it happened that disorder was on occasion represented by inversion, still it would not be explained just why reversal was the particular means resorted to. As for the postulated de-structuring of human nature, the formulation is cast in causal or at least sequential terms, but there is no evidence that the conceptual enterprise "leads" to reversal.

V

These prior investigations of reversal have done much to clear the ground, and prepare the way for an explanation, but they do not much succeed, I find, in arriving at a perspicuous account of the mode of relation posited as reversal.

In particular, they leave unanswered two main questions: (1) By what criteria does each of the kinds we have listed qualify as a reversal? (2) What have these various contrasts in common as reversals? In order to tackle these questions, let us first turn back to the list of kinds of contrast which ethnographers have commonly reported as instances of reversal.

We took special note above of the fact that "reversal" is a spatial metaphor (as also is "inversion"), and we posed the question of how apt this figurative term might be to the comparative analysis of social facts. Certainly it fits some kinds of reversal pretty well, namely those that are explicitly directional: these include upside-down, inside-out, backward constitution, backward action, and right/left. If suspect aliens to the Lugbara are upside-down, if the clothes of a Japanese are put on inside-out, if the Toraja dead do everything with the left hand, and so on, clearly what is imagined or performed agrees in some respect or another with a change of 180 degrees in direction and thereby effects a reversal in a spatial sense. But already there are differences to be remarked among even these cases.

In upside-down reversal, a physical object is rotated vertically so that its extremities are transposed. There is a difference in the case of inside-out reversal, in which what was inside is brought outside and conversely, so that what was concealed is revealed and the other way round; this is strictly an inversion, not a rotation, and it has a distinct outcome. Backward constitution is different yet again, in that what is normally at the back is imagined as having been brought to the front, which is neither a rotation nor strictly (like inside-out) an inversion. Backward action is different, in that in this case there need be no rotation or inversion, nor even a transposition, but there is instead some action that is characterized as being backward—and there are various ways in which this can be done. In the switch of preeminence from the right to the left, what is involved is a transposition of values and actions from one side to the other; this change can be conceived as directional, in that it is defined by reference to opposed points in a lateral dimension, but it is very different from each of the other spatial operations. Thus although these five kinds of reversal can all be classed as directional, they involve three different dimensions and a number of other differences besides.

The matter becomes more complicated in the case of transvestism, for with this case we decidedly enter the realm of what is usually called the metaphorical. The action is material, but it is not a rotation or other such operation in any dimension; it is the symbolic assumption of attributes of the opposite sex, and it is by virtue of the categorical contrast between the sexes, and the switch of their specific attributes, that the action is commonly reported, figuratively, as a reversal. The action can be carried out, moreover, by a member of only one sex, without a corresponding or reciprocal action by a member of the other sex. So in some instances transvestism resembles the transposition of lateral values, and in others it does not; in some instances it resembles an exchange of proper attributes, and in others it does not. These are variations of a formal nature, quite distinct from accidental dissimilarities in semantic content, and they make it hard to treat transvestism as a single distinct kind of reversal or to assimilate it to other kinds.

Lexical reversal qualifies fairly clearly as reversal, in that the order of the syllables at least is reversed; that is, their relative positions are transposed. But although we can recognize an order in a series of spoken sounds, it is only metaphorically that we can describe the sounds as occupying certain positions. (Literacy affects the issue, as also does the mechanism of sound recording, but not as a general factor in the comparative study of symbolism.) When the dead of the Toraja pronounce words backwards, the order in question cannot be presumed to be, in the view of the Toraja themselves, spatial, and so far as we can tell from the evidence it is not spatial; it pertains to a phonetic sequence.

When we come to semantic reversal, the operation is not formal but is defined precisely by content. Words and phrases are uttered with meanings that are the opposite of the normal, and it is in this regard that such linguistic features can be reported as reversals. How apt this description may be is dependent, not on a spatial interpretation of any kind, but on the pertinence of the notion of propositional reversals. We shall take up this topic in due course.

With those affronts to the customary which are taken as reversals, the grounds of comparison shift yet once more. If a mythical ancestor commits incest, or proffers a foot instead of a hand in greeting, he is doing something which makes an extreme contrast (most likely of a normally objectionable nature) to usual or correct behavior. The terms in which the extremity is assessed have to do, implicitly at any rate, with jural norms and canons of civility. None of these has an *a priori* reverse such as could be put into effect or be ascribed to other individuals, and the most we are free to assert is that certain wide differences are being posited or enacted. We cannot concede that they are reversals until we have a clearer notion of what is to count as a reversal—and it is with this very end in view that we are examining such differential judgements.

When we give up the criterion of extremity (another spatial metaphor, of course), and settle for mere difference, the case for reversal becomes still weaker. Although Kruyt states that the Toraja dead do everything "precisely the other way round" from the living, what he actually describes, and explicitly designates as such, are simple differences. If these differences have a peculiar significance in the eyes of the Toraja, this is presumably because they are symbols of the unseen world of the dead and other spirits. This is a traditional ascription of meaning, and not one which by any detached appreciation should induce us to class the differences as reversals. Indeed, Kruyt himself concludes by writing that a dead person among the Toraja is merely "a different kind" of being.

Lastly, so far as our list goes, there are the "inversions" reported from Roti which turn out to consist in not existing; that is, certain things which normally are done are just not done. There is admittedly a sense, in English usage, according to which it could be said that to abstain from doing something was the reverse of doing it; but it is only by exploiting this idiomatic recourse to the limit that abstentions can well be called reversals. Similar considerations apply to the sheer ignorance of sexual intercourse, kinship, and marriage among the first Lugbara.

Without adducing any more than the present twelve "kinds of" reversal, therefore, it can easily be appreciated that what ethnographers have readily described as reversals turn out to cover a range of more or less disparate phenomena. For that matter, we have already seen that Babcock actually defines as inversion not only any act that inverts or contradicts, but also any that abrogates or presents an alternative to the normal. What then can be the justification for classing together such a divarication of social facts?

It may help to emphasize this question if we keep in mind, as we proceed, a paradigm contrast: What is the resemblance, under the rubric of reversal, between (1) the image of a Kaguru witch walking upside-down on his hands and (2) the pronunciation of the word "lecture" syllabically backwards as "ture-lec"? If these are both reversals, as by ethnographic practice they can easily be counted, and are equally such, what exactly is the congruity between them?

Babcock's recourse, in summing up a selection of papers directed especially to the topic, is to use the phrase "symbolic inversion" as an organizing concept for diverse perspectives on

what she calls "cultural negation" (1978: 14, 32). This is a tempting move, for reversals can be interpreted as symbolic statements of cultural values, only in the form of negations; and, in
Ayer's words, "to say what things are not is itself a way of saying
what they are" (1954: 48). Moreover, there happens to be in traditional logic a means of obtaining an equivalent proposition, by
negation, which is actually called "inversion." But the very parallel tends to work against the pertinence of negation, for this latter process in logic is an operation that applies to propositions;
and it is only in a figurative sense that symbolic actions, such as
reversal, can be said to assert anything propositional.

Not only this, but the idea of negation is itself not clear or
certain, and logicians have long had a great deal of trouble with
it. Frege concluded that it appeared "impossible to state what
really is dissolved, split up, or separated by the act of negation"
(1960: 124). Wittgenstein at first postulated a "feeling . . . as if the
negation of a proposition had to make it true in a certain sense,
in order to negate it" (1953: sec. 447), which might appear somewhat to serve our present purpose; but he later said of negation,
as a "gesture" of exclusion or rejection, that it is "used in a great
variety of cases!" (sec. 550), and his treatment of certain illustra-
tive cases was highly circumstantial, not formal.

In any case, it would be only by a very forced use of the idea
that one could say that transvestism, lexical reversal, and abstention or ignorance were forms of negation. It is true that in
one analysis I have myself referred to the contrariant attributes
of a culture hero, Mpuga Rukidi, as "the very negation of all
valid credentials" (1973a: 327); but this phrase was an epitome
of a complex cultural confrontation, not a logical analysis of
any proposition or set of propositions capable of being denied.
It is in this expository guise, also, that Babcock's allusion to cultural negation can be conceded to have its proper force; strictly
speaking, it is rhetorical rather than demonstrative, persuasive
rather than analytical.

Nevertheless, to regard reversals under the aspect of negation can still procure some advantage. Ayer, in his discussion of
negation, introduces the idea of "complementary predicates."
Examples from ordinary speech (which will sound particularly
familiar to any student of symbolic classification) are "odd and
even," "light and dark," "wet and dry" (1954: 48). In the devel-

opment of his argument, Ayer reaches two conclusions: that any two statements are complementary if and only if each comprises in its range (sc. of application) the ranges of all the statements which are exclusive of the other (p. 54); and that two statements are complementary if and only if each is the negation of the other (p. 61). Notwithstanding the unavoidable reservations about considering symbolic representations as statements, these conclusions underline an important feature of reversals: namely, the feature of opposition.

Reversals are often described as implicating opposites, and some ten out of our twelve kinds of reversal can be defined by reference to opposition. The key to a perspicuous account of reversal might seem, then, to be available in this feature; but, as we have already remarked, the modes of opposition are numerous, and no one of them is to be regarded as basic or definitive.

Yet there is still a key of a sort—for since reversal is in the main inseparable from opposition, in some sense or another, it follows that there may be at least as many modes of reversal as there are of opposition. We have to say "at least," because of the variety in the modes of negation which can frame the oppositions, either those which are interpreted as implicit statements or those which have a propositional formulation, and because in principle any opposition can implicate more than one negation. When the dead of the Toraja understand "forwards" either as "backwards" or as "stand still," which as types of action are certainly different, it is a question whether both of the semantic reversals are equally opposite to "forwards," and also whether they are negations—and, if so, whether they are in the same mode.

There is a further consequence. It has just been asserted that reversal is inseparable from opposition; in some instances this is patent, in others it is at least arguable, and in some sense or other the assertion can be said to apply to most of the kinds of reversal in our list. Given so much, it might be contended that the answers to our two questions at the beginning of this section have been there all the time. Each of the kinds of reversal, it might be responded, can be assessed as such by reference to whether or not it implicates an opposition; and what the varied contrasts have in common, if they are to be counted as reversals in this sense, is the relation of opposition. These propositions entail that the possibility of discriminating a unitary class of reversals depends on the prior definition of what is to count as an opposition; but since there is no essential definition of the specific features of opposition, except by an arbitrary stipulation, it follows that there is no such definition of reversal either.

I think indeed, and on other grounds as well, that this is a correct conclusion, but it could not have been proposed as a premise to the present investigation: first, because an analysis of reversal could not be founded on an intrinsically disputable concept, namely that of opposition; second, because reversal had to be analyzed in its own circumstantial terms, as it has been variously reported by ethnographers and labeled by comparativists, before it could be correlated with another abstraction. As the investigation has worked out, however, the analysis of reversal has converged with that of opposition, and the character of each relation reflects that of the other. Neither is clear or certain, and neither is a constant by reference to which a monothetic class of social facts can be discriminated.

This does not mean that the notion of "reversal" is devoid of scientific use, but it does mean that it cannot have some of the uses that have been ascribed to it. What use it may still have as an analytical concept depends on a resumed analysis of its characteristic features.

VI

The phrase "characteristic features" indicates in itself a particular form of use. The class of reversals is polythetic, i.e., constituted by sporadic resemblances (Needham 1975 and chap. 3 above), and among the dyadic contrasts comprised in the class there is no specific feature that is common and essential to all instances of reversal.

We have already seen, in the previous section of the argument, that connections can be traced from one kind of reversal to another, and that the connecting feature repeatedly changes, so that the list of kinds makes up a Vygotskian chain complex in which there is no continuous or central significance. The uses of the spatial metaphor of reversal differ from case to case, and it is the figurative description which conduces to the assumption that there is an essential similarity among them. This assumption is incorrect and, to the extent that certain kinds of reversal have anything in common, this is primarily what is pictorially common to the spatial metaphor and its associated images. A commitment to the distinctive features of the metaphor itself, rather than to a scrutiny of its uses, leads to the idea that there is a paradigm case of "reversal," such as upside-down personages, and that other "kinds" of reversal are more or less valid replications of this. But an equal attention to the various social facts that are classed as reversals quickly undermines this presumption. There is no *a priori* form of reversal, as a type of social fact, and the instances of what are ordinarily described as reversals do not compose a monothetic class. This means that causal or correlative propositions about "reversal" can have only a qualified scientific value, if that.

But this is not the end of the matter, and there remains a further recourse in the examination of this concept. If what is at issue is whether or not the disparate kinds of reversal are in some respect isomorphic, we need to turn to formal analysis.

In this undertaking, the aim will be to formulate by abstraction the terms and transformations which are implicit in each of our twelve examples. The governing rule will be to introduce no distinction that is not required in order to represent the distinct form of the example; each formula will be reduced so far as practicable, without representing differences of content. The numbers under which the formulas are presented correspond to those of the examples listed in section III above; the notation will be explained as the analysis proceeds.

1. $(a, x) \Rightarrow (a, y)$. The constant a stands for the dimension, in this case verticality, by reference to which a variation is registered; the comma signifies adjunction; x and y stand for the contingent attitude of any given subject by reference to this dimension: e.g., either the right way up (x) or upside-down (y). (The converse ascriptions are formally identical with these.) The sign $\Rightarrow$ stands for transformation.

2. $(a, x) \cdot (b, y) \Rightarrow (a, y) \cdot (b, x)$. The constants stand in this case for the aspects of radial depth: e.g., a the outside, b the inside. The terms x and y stand for the contingent position of whatever is judged to be outside or inside by reference to this dimension: e.g., x outside, y inside.

3. $(a, x) \cdot (b, y) \Rightarrow (a, y) \cdot (b, x)$. The constants stand for limits on the dimension of horizontal depth: e.g., a the front, b the back. The variables x and y stand for objects (in the case cited, parts of the body) which are assigned to these points.

4. $(a, x) \Rightarrow (a, x^{-1})$. The constant a stands for a vector; the variables stand for action in agreement with this direction (x) and, to employ the mathematical sign for inversion, x^{-1} for action inverse to it.

5. $(a, x) \cdot (b, y) \Rightarrow (a, y) \cdot (b, x)$. The constants stand for right (a) and left (b); the variables x and y stand respectively for the contingent but opposed values differentially ascribed to them.

6. $(a, x) \cdot (b, y) \Rightarrow (a, y) \cdot (b, x)$. The constants stand for the sexes: e.g., a for male, b for female. The variables stand for the costume, appurtenances, posture, and so on, appropriate to each sex: e.g., x for masculine, y for feminine. In the case of nonreciprocal transvestism, this formula can be adjusted: e.g., $(a, x) \cdot (b, y) \Rightarrow (a, y) \cdot (b, y)$.

7. $(a, x) \cdot (b, y) \Rightarrow (a, y) \cdot (b, x)$. The constants a and b stand for jural categories: e.g., moieties or genealogically definable statuses; the variables x and y stand for members of other categories assigned to these categories.

8. $(a, b) \Rightarrow (b, a)$. The constants a and b stand for any significantly combinable syllables. There is no dimension or vector against which they are set; there is merely the temporal order in which they are conventionally uttered, and this order is reversed. In the case of first and last syllables, in a tri-syllabic word, the difference in the formula is inconsequential; namely $(a, \ldots, c) \Rightarrow (c, \ldots, a)$.

9. $(a, x) \Rightarrow (a, \sim x)$. The constant a stands for any significant word or phrase, and x for the significance conventionally ascribed to it; $\sim x$ stands for the negation of x or for an opposite meaning.

10. $(a, x) \Rightarrow (a, \sim x)$. The constant a stands for a distinct form of behavior; the variable x stands for the action of an individual or a class of individuals in conformity with that standard; $\sim x$, the negation of x, stands for action consistent with rejection of that standard.

11. $(a, x) \Rightarrow (a, \sim x)$. The constant a stands for any conventional form of action, and x for the contingent behavior of an actor with reference to the convention; in this case, $\sim x$ stands for not-x, i.e., behavior simply different from x.

12. $(a, x) \Rightarrow (a, \sim x)$. The constant a stands for a conventional form of action, and x for contingent behavior in accordance with the convention; in this case, $\sim x$ stands for the absence of x: e.g., abstention from or ignorance of the normal practice.

Before we estimate the cumulative impact of this formal analysis, there are certain qualifications to be registered.

The first is that although the notation resembles that of symbolic logic, the analysis is not logical; the discrimination of constants and variables is *ad hoc*, and the transformations are not inferences. Also, the notation employed is simple and convenient, but it is by no means the only one applicable; different notations, including for example matrices, could be used and might have other advantages.

The second qualification is that, in ascribing a particular form to a certain example, there is no implication that the example has only that one form. Some of the examples can be represented by more than one formula; where this is so, I have employed that which resorts to the fewer terms and other discriminations.

Thirdly, one formula or another may seem disputable, or a particular feature of the notation may be open to question, and the possibility of revisions in such regards has to be admitted in advance.

In general, then, the point to be taken is that the formal analysis presented above does not possess the mathematical kind of certainty or decidability that its notation appears to promise; and also that the formulas are not capable of deductive operations. Nevertheless, the exercise leads to results which seem decidedly useful.

The twelve kinds of reversal listed in section III were treated as relatively distinct one from another, and the possibility was mentioned that the same ethnographical instances might be distributed among yet more numerous kinds of reversal. Certainly a concentration on content would be likely to lead to more and more distinctions, whereas a concentration on form proves drastically to reduce their number.

The twelve kinds of reversal listed can be reduced to just five formulas:

i. $(a, x) \Rightarrow (a, y)$ [1]

ii. $(a, x) \cdot (b, y) \Rightarrow (a, y) \cdot (b, x)$ [2, 3, 5, 6, 7]

iii. $(a, x) \Rightarrow (a, x^{-1})$ [4]

iv. $(a, b) \Rightarrow (b, a)$ [8]

v. $(a, x) \Rightarrow (a, \sim x)$ [9, 10, 11, 12]

The first thing to notice is that, if we take the distinctive feature of reversal to be transposition, formulas *iii* and *v* are not reversals since, in them, all that is involved is the determination of a variable difference. This leaves three formulas of reversal as transposition (*i*, *ii*, and *iv*); these are not individually reducible, and they cannot regularly be reduced one to another, let alone to a single comprehensive formula.

Before we sum up the lessons of this section, there is an obvious matter—itself a formal feature—to be settled concerning the five basic formulas. Three of these are represented each by only one of the twelve kinds of reversal, and it is therefore the more necessary that their claims to autonomy should be looked into more carefully.

Formula *i* calls for special justification, for at first sight it looks as though it might be assimilable to formula *iii*. The question is why an opposite direction should be registered as *y* rather than as x^{-1} and thus as an inversion in the sense of the latter term. This is indeed a matter for rather fine judgement, but the first consideration that led to formula *i* as it stands is that *y* is an attribute (specifically an attitude of an object along a linear dimension), whereas x^{-1} represents an operation (specifically a directed movement along a vector). Another consideration is that along a linear (specifically, a vertical) dimension there are only two possible attitudes: one, by implicit definition, is the right way up; the other, occupying the only alternative attitude along this dimension, is upside-down.

In the case of formula *iii*, the peculiarity is that the action is directed but the posture is opposite to the direction. In any case, the methodological force of the formal analysis would persist even if formula *i* were subsumable under *iii* or vice versa; for in addition to the resultant formula (*i* + *iii*), there would still remain the three other formulas (*ii*, *iv*, *v*), and these are sufficient to reach the conclusions that I wish to stress.

It is true that among these latter formulas there is another instance, namely formula *iv*, in which the formal scheme is matched by only one example (8), but this is so distinct and is so paradigmatically a transposition that it does not need special justification on this score. More generally, in any event, even if the twelve kinds of reversal could actually be reduced to just two formulas, and if these could not be brought together under a single more fundamental formula of reversal, the main conclusions of the formal analysis would still stand.

In place of an intuitive apprehension of an array of similarities and differences among social facts, or even a polythetic analysis of a miscellaneous class, we now have a more perspicuous representation of the varied contrasts that have been conflated under the notion of "reversal." In place of the deceptive unity of a single absolute mode of relation, we have to cope with a variegation of reversals, as well as with contrasts that are hard to accept as reversals in any strict or useful sense.

VII

The outcome, then, and one which has been arrived at by two distinct methods, is that the notion of reversal does not denote a simple relation with a strict formal definition. To treat ethnographic reports of reversals and inversions as a monothetic class of social facts is to fall into a category mistake.

These conclusions might be resisted on the ground that alien ideologies seem to group together just such a variety of "reversals" as we started by examining, and that they appear to posit significant resemblances among them. Thus the Toraja make a set of contrasts between the living and the dead by resort to upside-down (1), backward constitution (3), lexical reversal (8), semantic reversal (9), right/left (5), and difference (11); alternatively, they can be said to exemplify all five of the basic formulas. Since all of these criteria are used by the Toraja themselves in characterizing the dead, and also other spirits, it might be inferred that the criteria must possess something in common.

But this is not analysis; it is to retreat to the very ethnographic reports which have seemed to call for analysis. And in any case the inference is invalid, for all that is shown by the convergent denotation of the dead by means of various reversals and differences is that these means of denotation serve precisely that end. What they have in common is that they denote the dead, not that as criteria they have anything in common among themselves. Also, there are other means of denoting the distinct character of the dead—e.g., insubstantiality and residence in an unseen world—but not all attributes of the dead qualify as reversed characteristics in any sense of reversal. It cannot be admitted therefore that the criteria in question must possess something in common, over and above their common denotata, but we have to look and see what exactly they do in fact have in common—and this compulsion takes us back to the very beginning of the present argument.

What then happens to comparative propositions about reversal, e.g., the suggestion that it is significantly correlated with boundaries? Clearly, if my re-analysis is right, they cannot any longer be taken as they stand, but each has to be recast or reinterpreted in more rigorous terms. It cannot be accepted, for instance, that reversal in general is an elementary possibility, "in a formal sense," in the symbolization of the ambiguous or suspended status of a subject at the marginal point between one category and another (cf. Needham 1979: 43).

Yet all the same, there are some modes of reversal which can readily be admitted as reversals—in one or another sense that is to be determined—and which are in fact widely employed in marking categorical boundaries. Their distribution and their common forms still lend support to the inference that reversal, as a means of marking a contrast, is not merely an imaginative possibility but also "a positive proclivity by which men tend to be influenced in their collective representations" (Needham 1978: 36). There are indications of this character in social usages which are not overtly categorical, let alone cosmological or mystical, and which are resorted to for amusement, as in folklore; and also in back-slang, i.e., saying each word more or less backwards as in "Uoy nac ees reh sreckin ginwosh" for "You can see her knickers showing" (Opie and Opie 1959: 320).

Psychologically, some kinds of reversal certainly appear to have an intrinsic appeal and impact, and professional students of the imagination would be of great service if they could help to locate the grounds of this efficacy. This may not be strictly necessary to the comparative study of social facts, but I take it that any globally distributed form of representation is correlated with general facts of individual cerebration and that an anthropological investigation in the widest sense is therefore incomplete until these latter factors have been sought out. Admitted, it is very difficult in principle to be at all sure that the expression of any idea or image is effectively spontaneous, though sometimes dreams and works of art seem to proffer evidence of a kind. I cannot in fact cite any convincing indications that reversal is an untaught proclivity of human beings in coming to terms with certain aspects of experience, but perhaps an example from modern fiction may hint at a case.

In *Le Roi des aulnes*, by Michel Tournier (1970), a little girl in a garage is struck by a flying blade from a car fan and is knocked unconscious. The protagonist, Abel Tiffauges, lifts her and is at once overcome by an unbearable and rending sweetness, a euphoric wave of beatitude. The objective experience registered is the girl's weight, and it is this that provokes his bliss. Reflecting on the heaviness of their two bodies together, Tiffauges concludes that a kind of levitation has been produced in him by an added weight: "An astonishing paradox!" he writes. "The word *inversion* at once presents itself under my pen. What happened was in a way a change of sign: plus becomes minus, and the other way round. A benign, beneficent, divine inversion. . ." (p. 133; original emphasis).

I am not sure what can be made of this instance other than simply to observe that in this case an imaginative writer procures a particular (if Humbertian) effect in his hero, as also possibly in the reader, by phrasing the poignant paradox as an inversion. Perhaps a lead is provided by the stock expression about change of sign. In an earlier discussion of reversal, I have found myself writing that "a symbolic classification is given an intensified application by reversing the signs, as it were, on the values of its categories" (1979: 41). This is an assertion which now will not bear much in the way of close scrutiny, but the allusion to the operation of reversing the signs does call to mind the technical uses of inversion in mathematics, logic, and music; also the reversals played upon in a well-known painting by Magritte and in the tessellations of Escher, and we know what an appeal these images have had in book-jacket designs and in prints on students' walls.

In analyzing certain uses of the verb "to see," Wittgenstein considers the experience of noticing an aspect; and in examining the paradoxical experience of "change of aspect," he adduces the familiar figures of the duck–rabbit and the step (1953: 193–94, 203). In both of these cases an effect is procured by a reversal: the duck looks to the left, then the rabbit looks to the right; the step is now convex, now concave. And in touching specifically on reversals, Wittgenstein finds a "different difference" (as the English translation felicitously renders "ein anderer Unterschied") between one example of reversal and another (p. 198).

Perhaps, after all, the critical task of the comparativist is done when he has elicited the changes of aspect that are to be discerned in the interpretation of reversals.

Parallels or like relations
alternately releeve each other.
Sir Thomas Browne

6

Alternation

I

There are two distinct kinds of interest to be found in the structural analysis of social facts: that is, in the employment of relational abstractions in the scrutiny of collective representations and forms of social action.

One kind of interest stems from the remarkable utility of such analysis in the determination of systems. The other resides in the inference that the analytical notions thus relied upon, these being taken as common to the analyst and to the phenomena under study, are indexes to common properties of the human mind; in other words, that in the construction of social systems the fabricators were tacitly guided by the very relations which the analyst makes explicit.

Examples of concepts possessing these two kinds of intellectual value, analytical and cognitive, are opposition, analogy, and homology. These are very general in their scope, and this generality is matched by their open texture and their power of accommodation (cf. Needham 1980a, chap. 2). Although they are fairly steadily recognizable, just as they are discrete one from another, they can inform an illimitable variety of social forms and can be conceived figuratively as being exceedingly resourceful. These particular relations are patent features of collective ideation, and by deeper analysis they can be shown also to characterize systems of social action; thus they commonly organize cosmologies and other forms of symbolic classification, and they can be demonstrated to underlie such different institutions as funerary prestations, prescriptive alliance, and complementary governance.

Other relational concepts are more obviously characteristic of forms of society, and they can often be expressed in jural terms; examples are symmetry, asymmetry, and transitivity. These concepts are far less general in their scope than is the former set of relations, but at the same time they are much clearer and they can be given satisfactory formal expressions. The degree to which they can be accorded a necessary character (the ultimate formal property) varies from one concept to another, yet all the same these relations have a clear claim to be considered basic, in some sense, and to the comparativist at any rate they are practically indispensable. Our attention is drawn to them not merely by philosophical tradition but by their general uses in comparative analysis, and on this ground we attribute to them a global "distribution" among systems of social facts. They are of course formal abstractions, i.e., constructs in a particular epistemological tradition, and as such they are not precisely replicated as verbal concepts in other languages; but on occasion there do exist vernacular equivalents, and these tend to confirm the universal character of the underlying relations. In any event, it is consistent with their subliminal nature, and with what I take to be their undeliberative origin in the unconscious, that in the main they are not made explicit by those whose collective representations are ordered by them.

Even if it should not be conceded that opposition, symmetry, and the rest correspond (as I think they do) to natural proclivities of thought and imagination, their pragmatic utility in empirical analysis makes them objects of outstanding interest. Granted only this much, the question is: How many concepts of the kind have we to recognize? It is not to be presumed, after all, that the half-dozen relations that I have cited are all that may be at work, especially since it is only relatively recently that they have been so methodically isolated in the comparative study of social facts. Also, some of even these relations

have not yet been adequately explicated or given a suitably formal expression. On the other hand, we do not ask if the present
set of concepts makes up a "full" list of formal relations of the
dual nature in question, for this query would presuppose a
limit of completeness which is itself disputable and which we
have no prior reason to suppose can be attained.

Nevertheless, we do not suppose, either, that there are no
further relations to be discovered (or contrived) in addition to
the present list. Taken merely as technical resources, the present
instances prompt the search for further relations which will
match them in their fundamental character and thereby in their
dual interest as indexes to constant proclivities of the mind and
as tacit guides in the fabrication of social systems.

The candidate that I now wish to consider, as a basic concept
of this kind, is "alternation."

II

A visual example may set the scene. In his monograph on symmetry, Weyl depicts a frieze of Persian bowmen
from the palace of Darius at Susa, dating from possibly the 5th
century B.C. It represents a file of armed men marching toward
proper right; in form they are identical, and Weyl is concerned
to point out that the frieze is thereby an example of pure translation. But "the basic translation covers twice the distance from
man to man because the costumes of the bowmen alternate"
(Weyl 1952: 48). The robe of one man is ornamented with circular patterns, then that of the next with a rectangular design,
and that of his successor with circular again, and so on. Thus
the formal property of translation is qualified by an alternation. This is not a detail to be registered without query, nor is it
a mere artistic quirk on the part of the architect. It is one of the
earliest surviving examples of recourse to a mode of relation
which organizes a great many other things.

Here is an introductory example from the nonvisual field
of social forms. The Meru of Kenya are organized into age-
classes, and these fall under "an alternating dual division"
(Bernardi 1959: 21). These divisions are named, and the age-
class in power belongs to each of these divisions in turn; it is

"always of the opposite alternating division from the next preceding and the next following age-class in power" (p. 22). What the new age-class receives is "a kind of delegated power by which its members are expected to assume a primary and thorough control of things communal" (p. 23).

The occasion of handing over power from one age-class to the next is a ceremony called *ntuiko*, meaning "break" (p. 22). This covers a long period during which all the country is happily feasting: "the elders of the retiring age-class are at liberty, during that period, to do whatever they like" (p. 23); they spend their time going around to drinking parties and "indulging in the sort of behavior that would be considered shameful at any other time" (p. 24). At a special ceremony, the new age-class is blessed and given a name; both of these functions are exclusive privileges of a dignitary called the Mugwe. In the Igembe subtribe, "the election of the Mugwe follows the two alternating divisions" (p. 33); in the Igoki subtribe also, "the election and retirement of the Mugwe is closely related to the formation of the age-classes and hence with the alternating divisions of the age-system" (p. 37). It would be a rather lengthy business to demonstrate how the age-class system and the office of the Mugwe are integral to the constitution of Meru society (see Needham 1960a), but, on the assumption that they are, the essential point to take is that the age-class in power comes alternately from each of the two named divisions and that the Mugwe himself is elected from each of these divisions alternately. Among the determinants of the form of Meru society, a crucial and definitive part is thus played by the relation of alternation.

We shall take up presently (sec. V below) a number of other examples from the comparative study of social forms; but before we do that, we should look at the definition of this mode of relation which can characterize equally the pattern of a frieze and the constitution of a society.

III

A dictionary definition of "alternation" is that it is "the action of two things succeeding each other by turns."

The substantive notion is based on the verb "to alternate," which is derived from the Latin *alternātus*, past participle of *alternāre*, to do things by turns, from *alternus*, from *alter*, one or the other of two, second (Onions 1966: 29, s.v.).

The *Oxford English Dictionary* distinguishes two applications of the adjectival form. First, it is said of things of two kinds, so arranged that one thing of one kind always succeeds, and is in turn succeeded by, one of the other kind; thus

$$* \quad \dagger \quad * \quad \dagger \quad * \quad \dagger \quad * \quad \dagger \quad . \quad . \quad .$$

occurring by turns, as alternate day and night, etc. Second, it is said of things of the same kind taken in two numerical sets, so that one member of each set always succeeds one of the other; thus

1   2   3   4   5   6   7   8   9   10   11   12

so that 1, 3, 5, 7, 9, 11 make up one set, and 2, 4, 6, 8, 10, 12 make up the other set (*O.E.D.* 1971, s.v. "alternate," adj., senses A.1 and A.3). We shall see as we proceed whether this distinction is of consequence in the practice of the comparative analysis of social facts.

Another preliminary, however, is to seek equivalent expressions in other languages. Whenever we are investigating a verbal concept, and especially one that we suspect may correspond to a basic operation of thought, it is advisable if not imperative to see if other linguistic traditions provide for the recognition of anything resembling that concept (cf. Needham 1972, chap. 3). The Romance languages do not make a sufficient contrast in the present case, by reason of the latinate derivation of "alternation," and our first recourse therefore is to look at the German. In this language the equivalent is *Abwechslung*, from the noun *Wechsel*, change, exchange; this is explained as meaning essentially to give ground or yield, to make way for (Grebe 1963: 756, s.v.), and thus as developing into such meanings as exchange, alternation, series. Similar considerations attach to the Dutch *afwisselen*, to alternate, from *wisselen*, to exchange or change; and like connections can be traced in other Germanic languages.

If we look up "alternation" in an English–Sanskrit dictionary, the nearest to an equivalent is *parivṛtti* (Apte 1914: 11, s.v. "alternate"). This has among its meanings the ideas of return, exchange, barter. But it also denotes: turning; moving to and fro; end, termination; (in rhetoric) a kind of figure in which one thing is represented as exchanged with another; substitution of one word for another without change of sense (Monier-Williams 1956: 601, s.v. *pari-*).

In Chinese, finally, there is a character conventionally romanized in the Wade system as *tieh* and translated as "alternately; to alternate." Karlgren supplies as further glosses: rotate; change, succeed, successively, by turns, repeatedly (1923: 259, char. 880). In addition to the radical, the character includes *shih*, which Karlgren glosses as: to drop, to lose; miss, neglect; to err, a fault. The composition of this latter character itself is explained as "to drop from the hand."

Even this brief comparative survey seems to show that in an array of disparate languages there are words which are to some degree equivalent to "alternation," and also that these words convey in common a complex of senses that includes most prominently the notions of change, substitution, series. The Chinese character is particularly significant in its intimation of omission or passing over. These comparisons give encouraging indications that what we call "the" idea of alternation manifests a natural tendency in the discrimination of relationships among things, and that it may thus constitute a basic concept of the kind we have in view.

IV

Yet "alternation" is not a technical term in formal logic or in other philosophical enterprises; it has no place in dictionaries and encyclopedias of philosophy. Nor is it a term with any general use in mathematics or in such exact sciences as physics and chemistry.

There is indeed in physics the term "alternating current," but this is merely a particular designation for a flow of current that periodically "reverses." What this verb in its turn means in this context is that a flow of electric charge starts from, say, zero, increases to a maximum, decreases to zero, reverses, reaches a maximum in the opposite direction, returns again to the original value, and repeats this process indefinitely. In other words, the phenomenon in question undergoes a periodic fluctuation which, in a standard description, constitutes a cycle. The word "alternating" here is no more than a convenient label; it does not indicate that the concept of alternation has a theoretical value in the study of electromagnetism.

Similarly, there is in botany the term "alternation of generations," designating a process of reproducing alternately by sexual and nonsexual means. In alternate generations, the individual born from an egg never assumes through a succession of transformations the character of its parent, but produces, either by internal or external budding or else by division, a number of new individuals; and it is this progeny of the individuals born from eggs which grows and assumes again the characters of the egg-laying individuals. These transformations are "the successive terms of a cycle, . . . a well-regulated cycle ever returning to its own type" (Agassiz 1859: 135–40). Although the phenomenon occurs in all multicellular plants, the term "alternation" is a convenient description but not a theoretical concept. Moreover, the equivalent description of "cycle" makes a significant divergence, as we shall see, from the kind of alternation that is commonly found in social institutions.

In fact, the first point of general consequence to be registered is that among collective representations, by contrast with natural phenomena, alternation does appear as a distinct mode of relation and in a wide variety of social forms. It is not to be defined by correlation with any particular type of institution, whether ideological or pragmatic, but for theoretical purposes it should be abstracted and formalized.

An adequate verbal formulation, to begin with, is that alternation is given by a sequence composed of two terms, in which neither term succeeds itself: e.g., $a, b, a, b, \ldots$ The constituent relation can be formalized as follows. Let e_i stand for a sequence of events (e), the index i stating that e can take different values. Then alternation is represented by the equation

$$e_n = e_{n+2}$$

It is to be understood that n can occupy any position in the sequence. Thus in the interrupted sequence $a, -, a, -, \ldots$, the alternative term *b* must occupy the vacant positions between the incidences of *a*; and likewise in *b*, −, *b*, −, . . ., the alternative term *a* must occupy the vacant positions between the incidences of *b*—so that *n* may be represented by either *a* or *b*.

The relation of alternation, considered thus far, is simple to apprehend; it can be adequately represented by a certain sequence; and it can be formalized by a simple equation.

The problem now is to account for the common occurrence of alternation as an organizing principle among social facts, such as the alternating divisions of the Meru age-classes and the parallel alternation in the election of the Mugwe. To this end, we need to expand the range of ethnographic cases in other societies and in institutions of various kinds.

V

Indonesia

Among the Makassarese and the Buginese, it is reported that when a marriage took place between individuals of unequal rank the resulting children used to be divided between the parents: the first-born "belonged" to the mother, the second to the father, the third to the mother, the fourth to the father, and so on (Wilken 1912, 1: 362). On Sumbawa, when a man of a certain aristocratic class married a woman of the commoner class, and in the event that no bride-wealth was paid, the children were divided: the eldest was assigned to the father's class, the second to the mother's, and so on (Wilken 1912, 2: 212–13). In these marginal marriages, therefore, a rule of alternating affiliation was brought into play.

Penan

The Penan of Borneo, like many other peoples of the interior, employ a system of death-names and teknonyms. There are a number of death-names, each referring to the relationship that obtained between the deceased and the survivor who assumes the name: thus if the grandfather dies, his grandson is called Tupeu; if the father dies, his son is called Uyau. Until the survivor is married, he or she may assume a number of such names in succession; but after marriage and the birth of children another naming practice comes into operation. After the birth of the first child the parents adopt teknonyms referring to the child: Tama N., father of N.; Tinen N., mother of N. If this child, or any subsequent child, dies, the teknonym is abandoned and the parent assumes a death-name referring to the birth-order of the deceased child: e.g., Uyung, first-born child dead; Akem, sixth-born child dead, etc. This latter death-name is kept until the birth of another child, in which event the parents will assume teknonyms referring to this new child. Among unmarried siblings there is a similar alternation, between personal name and death-name, occasioned by the successive births and deaths of brothers and sisters.

The particular usages and their interconnections are rather complicated, but the principle behind certain changes is simple enough: within certain degrees of relationship there is an alternation between (1) personal names and teknonyms occasioned by births and (2) death-names occasioned by deaths (Needham 1954; 1971c: 207).

Kariera

This is a classic example of the four-section system. The society is divided into four named sections: Banaka, Burung, Karimera, Palyeri. Every individual, whether man or woman, belongs by birth to one or another of these sections. Marriage is so regulated that Banaka is allied categorically with Burung, and Karimera with Palyeri. A child, of either sex, belongs by birth to the section of neither its father nor its mother, but to one determined as in this diagram:

$$\left[\begin{array}{l} \text{Banaka} = \text{Burung} \\ \text{Karimera} = \text{Palyeri} \end{array} \right]$$

The equation signs stand for marriage, and the vertical brackets indicate section assignment. Thus if a Banaka man marries a Burung woman, the children are Palyeri; if a Palyeri man marries a Karimera woman, the children are Banaka; and so on.

Now marriage is also regulated by a two-line terminology of prescriptive alliance, and hypothetically the males of each local descent group will be designated sequentially by the categories of one of the terminological lines. Hence there are in principle two kinds of local group, distinguished internally by section names and by relationship terms. The men of one categorical group belong in succession, by generations, to the sections Banaka–Palyeri–Banaka–Palyeri–, and so on; those of the other categorical group belong to Burung–Karimera–Burung–Karimera–, and so on (cf. Needham 1974a, chap. 3). There is no need to expatiate on the importance of the sections, which is cosmological and ritual as well as jural. The point here is that section-assignment in each of the categorical groups is ordered by alternation.

Aranda

The Aranda of central Australia are a paradigm case of an eight-section system. The structure of their relationship terminology can be represented as four patrilines articulated by a symmetric prescription, as in figure 1. The interest of this form

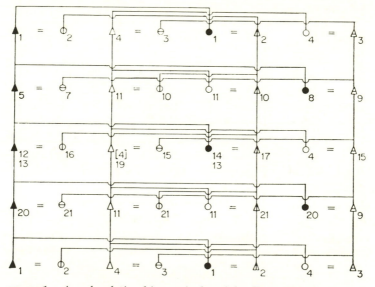

FIGURE 1.   *Aranda relationship terminology (after Korn 1973: 30, fig. 4).*

of classification, for the present purpose, is that any line gives wives to two other lines in alternation, and takes wives from these same lines in alternation. There is no direct or reciprocal exchange of women between any two lines, but for a given line in relation to an affinal line there is an intercalation of transactions with the other affinal line.

This process can easily be seen by concentrating on one line in the figure, e.g., the line of masculine symbols (triangles) in solid black. In the top genealogical level this line takes a wife (to translate the matter into a real transaction) from the line distinguished by a vertical bar; in the next, it takes from the line distinguished by a horizontal bar; in the level after this, a vertical bar again; then once more a horizontal; and finally a vertical. If the eye travels down the feminine symbols (circles) joined by equation signs to the solid black symbols, it is quickly seen that there is a regular alternation of two wife-giving lines. The same

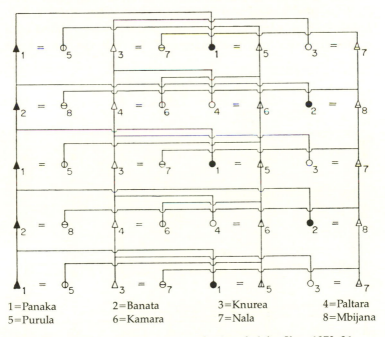

1=Panaka   2=Banata   3=Knurea   4=Paltara
5=Purula   6=Kamara   7=Nala   8=Mbijana

FIGURE 2.   Section-assignment among the Aranda (after Korn 1973: 31, fig. 5).

alternation can be demonstrated by tracing the zigzag distribution of the solid black feminine symbols as these are transmitted to each of the two affinal lines in turn.

A concomitant alternation can be seen in the pattern of section-assignment, as in figure 2. The members of the patriline in
solid black, for example, belong by turn in successive genealogical levels to the sections 1–2–1–2–1; the wives of these
members come respectively from sections 5–8–5–8–5.

In each of these schemes of classification, therefore, Aranda
society exhibits a fundamental and regular alternation.

Dieri

It is at least technically interesting that among the Dieri of
Australia there is a relationship terminology structurally identical with that of the Aranda, except that it is based on matrilines instead of patrilines (fig. 3). Also, there is no corresponding system of sections.

Despite these differences from the Aranda, the alternation

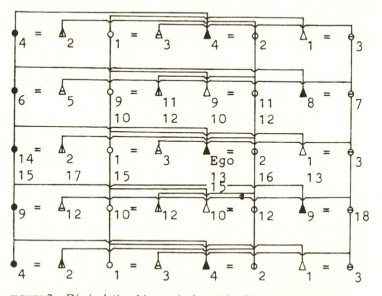

FIGURE 3.   *Dieri relationship terminology (after Korn 1973: 65, fig. 12).*

that expresses an essential aspect of this type of system can be clearly seen.

Iatmül

This society, on the Sepik river in New Guinea, offers a remarkable combination of various expressions of alternation. I shall base this example on Korn's exemplary analysis (1973, chap. 5) of Bateson's monograph (1936).

The structure of the relationship terminology consists of five patrilines articulated by an asymmetric prescription; any given line takes wives in alternation from two other lines, and gives wives in alternation to two further lines (fig. 4). The ethnographic information does not permit us to establish connections among all of the statuses implicated, but the mode of articulation emerges clearly from those that can be determined. Figure 5 depicts an integral system of asymmetric prescriptive alliance with five patrilines and alternation by genealogical level, a scheme which permits an easier recognition of the systematic significance of alternation in this sphere of Iatmül social life.

There are between 50 and 100 totemic patrilineal clans, and each of these is divided into *mbapma* (the literal translation of

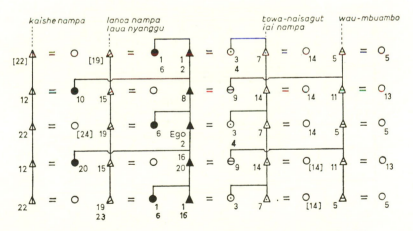

FIGURE 4.   *Iatmül relationship terminology (after Korn 1973: 92, fig. 22).*

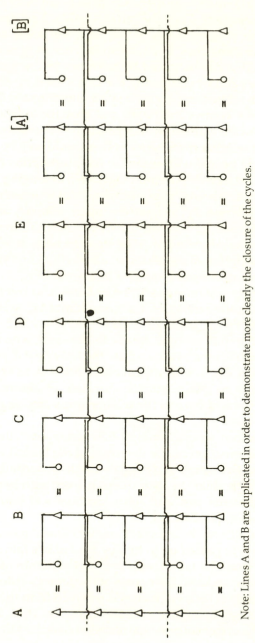

Note: Lines A and B are duplicated in order to demonstrate more clearly the closure of the cycles.

FIGURE 5. *Asymmetric prescriptive system with five patrilines and alternation by genealogical level (after Korn 1973: 97, fig. 24).*

which is "line") containing members of the clan belonging to alternate generations. Thus the father's father, Ego, and son's son form one *mbapma*, while father and son form another *mbapma* of the same clan. The correspondence of generations in different clans can be traced by reference to the system of totemic names and by comparing the terms used by different individuals in addressing a certain spirit. Bateson says that the *mbapma* do not control marriage, and that the relationship terms used toward women are independent of the *mbapma*. But Korn has argued that the *mbapma* system could well be useful in categorizing people with regard to the prescription that articulates the relationship terminology. Men belonging to one *mbapma* of a clan should marry women from a different clan from those married by men of the other *mbapma*; and the women of alternate generations could be easily discriminated in relation to the *mbapma*. It may be stressed here, at any rate, that the prescription necessitates an alternation, and that the constitution of an *mbapma* is defined by alternation.

The initiatory system is described by Bateson as constituted by two cross-cutting moieties. Each quadrant (Ax, Ay, Bx, By) is divided into three named generation groups—1, 3, 5, or 2, 4, 6—such that men in group 1 are the fathers of 3, who are the fathers of 5, while men in group 2 are the fathers of 4, who are the fathers of 6. The whole system of initiation is, as Korn observes, difficult to grasp; but the principle of alternation is nonetheless evident here in the relationships among generation groups.

In the relations among siblings, Bateson reports, the same principle of alternation can be detected. In the Iatmül vocabulary there are five terms to designate the first, second, third, fourth, and fifth child. In their quarrels over patrimony it is "expected" that the first and third brothers will join forces against the second and fourth.

An individual may bear a great many personal names, and most of them are passed on to him by his father's father; they are received from the father, who applies the names of his father to his sons, and also the names of the father's sister to his daughter. Another series of names is received from the mother's clan, which gives them to its sister's children. Thus, as Korn concludes, the system of personal names shows the same traits as the relationship terms and gives another clue to an understanding of the alternation by genealogical levels in the affinal system. It seems to reinforce the fact that an individual is linked with both his father's and his mother's clans, and thus should marry in a third one; "the alternation of personal names in one's own clan seems to indicate which is this third clan" (Korn 1973: 100).

Lastly, in the present treatment of this fascinating society, there is the topic of flute music. Among the Iatmül the flute is always a dual instrument. Two men play together, and their flutes are tuned by trimming the length so that flute A is exactly one tone higher in pitch than flute B. Then the harmonics of A will to a great extent fill the gaps in B's scale, "an arrangement which would seem to be a characteristic product of Iatmül thought" (Bateson 1935: 159). The two men blow in turn, moreover, and each man inhales while the other is blowing; they play a first phrase and then a second, and after this they revert to the first before proceeding to other phrases (pp. 160–61). Thus the flutes, which are "one of the most important threads in the whole fabric of the culture" (p. 164), also exemplify the principle of alternation which is so characteristic of Iatmül collective representations.

Age-group systems

In his comprehensive study of the fundamentals of age-group systems, Frank Stewart distinguishes a common type of intergroup relationship as "alternate linking" (1977: 125). In these cases there is some particularly close relationship between group S and groups S + 2 and S − 2; this may be associated with hostility between S and groups S + 1 and S − 1.

Stewart writes that alternate linking has been noted among the Iatmül, "though the details remain obscure" (p. 126), and this society may not have an age-group system of the type that is the subject of Stewart's monograph. As for alternate linking itself, it has been reported without the hostility between adjacent age-groups, but the close connection between alternate groups is definitive. Examples of this type of age-group system are reported from as far afield as East Africa, West Africa,

South America, and perhaps North America (p. 127), and self-evidently they are based on the principle of alternation.

VI

In particular contexts, alternation has been isolated for examination by Korn and by Stewart. It has also been the subject of more ambitious observations by Lévi-Strauss.

Korn, in a chapter entitled "Some Comments on Alternation" (1973, chap. 6), collocates the systems of social classification of the Aranda, Dieri, Mara (with four-line terminologies), and the Iatmül. These can be classed together, she writes, by some of the basic principles that they apply: they are all lineal; each of them presents the features of a closed classification; and in each of them the prescribed category can be defined by the relations of affinal alliance of one line with respect to two other lines. For each line, the principle involved in the determination of the prescribed category can be represented as is shown in figure 6.

Patrilines   Matrilines

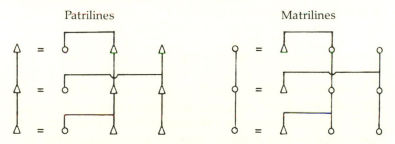

FIGURE 6.   *Structural module in alternating prescriptive systems (after Korn 1973: 121, fig. 29).*

Consecutive positions in the line of reference are related affinally to two different lines, and the number of lines in the system derives from the principle of symmetry or of asymmetry in force. "The alternation of terms in each line can then be seen as a result of three factors: (i) a lineal classification, (ii) a closed classification, and (iii) a principle according to which a line cannot be affinally related to the same line in two consecutive genealogical positions" (p. 121).

This "negative factor (iii)," Korn continues, can be seen as a matrilineal principle applied to a terminology composed of patrilines, or as a patrilineal principle applied to a terminology composed of matrilines. Neither of these ancillary principles can satisfactorily be subsumed under the notion of "double descent"; and the idea that systems of alternating generations are to be explained by intermarriage between sections is certainly wrong, since sections do not exist among the Mara any more than they exist among the Dieri, yet these systems are no different from that of the (eight-section) Aranda (p. 122). Instead, the systems of the Aranda, Dieri, Mara, and Iatmül can be defined by a principle such as that represented in figure 6. "Whether patrilineal or matrilineal, symmetric or asymmetric, with sections or without, this mode of articulation is present in all of them" (p. 123).

These arguments are entirely cogent in the special context of Korn's analyses—namely, a critique of the views of Lévi-Strauss on kinship—and up to a certain point they are satisfying also as a consideration of the principle of alternation in prescriptive systems. Yet even these proofs, decisive though they are for the theoretical issues directly in question, do not account for the phenomenon of alternation when it is considered more radically. At a later point in her monograph, Korn writes that the alternation of generations is not a rule, but "a distinctive feature (itself the resultant of certain principles . . .) of a type of terminology" (p. 136).

If we go back to those principles, however, we find that the relevant one is the third: namely, that the prescribed category can be defined by the relations of alliance of one line with respect to two other lines. Yet these relations themselves are defined precisely by alternation. As for the alternation of terms, which discriminates the genealogical levels at which alternation of alliance occurs, the third factor is a principle according to which a line cannot be affinally related to the same line in two consecutive genealogical positions. Now if this prohibition is not a rule, governing empirically the marriages contracted between lineal descent groups, what exactly is the principle except that of alternation? Certainly alternation by genealogical level is a distinctive feature of a type of terminology, but this feature cannot well be a resultant of principles which include alternation itself.

Under these aspects, then, an additional force in Korn's argument is that it compels us to reconsider more intently the grounds of alternation, as a mode of relation, and its claims to be accepted as a principle.

Frank Stewart's treatment of this relation, in the form of alternate linking in age-group systems, calls for only brief and less technical notice. He writes: "Alternate linking seems to me rather a mysterious phenomenon, and I can do little more here than say why some of the explanations for it that might come to mind are wrong" (1977: 125). One such explanation is by recourse to diffusion, which is indeed quite unconvincing. Another is by correlation with what is called paternal or fraternal linking, which Stewart can also show not to be well founded. A third is that alternate linking produces a division of age-groups in a sequence into two groupings or "streams." Certainly, also, a psychological explanation based on the Radcliffe-Brownian premise that there is a natural affective distance between adjacent age-groups does not serve, for Stewart finds that alternate linking has been reported without such hostility, just as the hostility between adjacent age-groups has been reported without alternate linking. In any event, Stewart deals with alternate linking in age-group systems, not with the general grounds of alternation, and even in his own field of evidence he finds the phenomenon to be rather mysterious.

One person, however, who has ventured to say something general about alternation is Lévi-Strauss. In a discussion of Frazer's views on the evolution of Australian section-systems by means of a deliberate and repeated bisection, he contends that the transition from nature to culture is determined by man's ability to think of biological relationships as systems of oppositions. These include an opposition between "the consecutive series (composed of individuals of the same sex) and the alternate series (where the sex changes in passing from one individual to another)" (Lévi-Strauss 1969: 136). It could be said, he continues, that human societies tend automatically and unconsciously to disintegrate, along rigid mathematical lines, into exactly symmetrical units. Then comes the crucial passage:

> But perhaps it must be acknowledged that duality, *alternation*, opposition and symmetry, whether presented in definite forms or in imprecise forms, are not so much matters to be explained, as [they are] basic and immediate data of mental and social real-

ity which should be the starting-point of any attempt at explanation (p. 136; emphasis added).

This is a bold (or at least a hazardous) suggestion, but it remains no more than that. At any rate, Lévi-Strauss's other allusions to alternation in the same work do nothing to clarify the concept.[1]

In discussing the process of segmentation from moieties to subsections, Lévi-Strauss asserts that harmonic regimes can acquire an autonomous structure only when they reach the stage of generalized exchange with n sections; prior to this stage the continuity is concealed within dual organization, while "later it is corrupted and distorted (as in the Murngin system) by the inevitable contamination of the principle of alternation" (p. 441). These propositions cannot be correct, since they repose on the distinction between harmonic and dysharmonic regimes; and Korn has demonstrated conclusively that Lévi-Strauss's inferences from this disputable distinction are fallacious (Korn 1973, chap. 2). Moreover, if we look at Lévi-Strauss's chapter on the Murngin system (1969, chap. 12), it turns out that the only pertinence of alternation resides in the contrast between a "regular" system of affinal alliance and an "alternate" system, which does not accord with our theoretical interest in a principle of alternation; all it means is that if the Murngin for some reason do not marry according to their proper rules, they turn instead to an alternative system.

At another place, Lévi-Strauss asserts that "the quartet of marriage with the matrilateral cross-cousin is the systematic application, to all degrees of kinship, of the formal alternation of sex on which the existence of cross-cousins depends" (1969: 444). This assertion rests on two mistakes. First, Lévi-Strauss relies on an old-fashioned and misleading genealogical diagram (p. 443, fig. 84; here fig. 7a) to represent the relationship to the "mother's brother's daughter," alias the matrilateral cross-cousin. In the first place, it is not a quartet but a quintet; Lévi-

---

1. To similar effect, Dumont, in discussing "alternating generations" in Australia, sees their distinctive characters as "aspects of a universal tendency," and he even risks the evolutionary surmise that "alternating generations are more primitive than a continuous flow of generations" (Dumont 1966: 238). Later, he emphasizes this alternation as "a basic feature, not to be reduced to others" (p. 249).

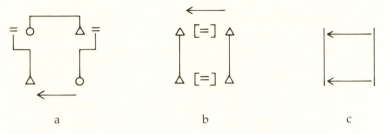

FIGURE 7.   *Diagrams of matrilateral cross-cousin marriage.*

Strauss has left out the necessary position of the father, and to include him raises the number of persons implicated to at least five. Then there is no need to plot feminine positions in order to represent the structure of the situation. If we employ Dumont's convention of a bracketed equation-sign to indicate affinal alliance (here fig. 7b), with a diacritical arrow to indicate wife-givers and wife-takers, there is no need to put a genealogical construction on asymmetric ("matrilateral") alliance. Nor is there any need, hence, to impute to the affinal relationship the "alternation of sex" which Lévi-Strauss apparently reads off from the genealogical specification M(f.)B(m.)D(f.). The reference to any female relatives is elided, yet the structure persists. This is even more evident if we resort to an alternative diagram (fig. 7c), which is at once more economical and more directly informative.

If it is responded that genealogical relationships can nonetheless be read from the present diagram (which is correct, as also for that matter they can be read off a matrix), the objection remains that there is no justification for treating genealogy and sex as though they were primary determinants of the structure. Also, much depends on just how a genealogical relationship is traced. For instance, if the emphasis is placed on a jural dominance of men over women, and if the purpose of the diagram is to convey the distinctive features of a social situation, then the premise that men treat directly with other men in the disposition of women (and, let us say, within the same genealogical level) could well be recognized in the specification of connections. Thus the relationship MBD might be rendered accordingly as FWBSZ, a specification which agrees with the premise but does not exhibit a regular alternation of sex.

Whether it would be appropriate to do so would depend on

the facts of the particular case, but another objection to Lévi-Strauss's assertion is completely general. He writes that the formal alternation of sex is applied "to all degrees of kinship"; but if we go only so far as the second cross-cousin, we find that a woman at the same structural location as MBD can be FMBSD, a specification that does not exhibit a regular alternation of sex. Admitted, there is a change of sex, as it were, at FM- and at -MB-, but this "alternation" is blocked at the same-sex adjunction -BS-. If this limited alternation (if that is what it is) is what Lévi-Strauss means, it is not what he says. And in any case, the formulation really means no more than that any genealogical connection between two patrilines (for example) that are affinally related must include, at some nonterminal point within the specification, at least one feminine term.

Lastly, Lévi-Strauss concludes that "bilateral marriage has the characteristic of alternation in common with patrilateral marriage" (p. 464). It is well known, of course, as Fortune demonstrated in 1933, that hypothetically a patrilateral prescriptive system would entail a perpetual alternation, from one genealogical level to the next, of the direction in which women would be transferred. But it is a hard matter to conjecture (since Lévi-Strauss provides no argument) what could be meant by a system of bilateral marriage, or symmetric alliance, that was characterized by alternation.

One interpretation might be that the characteristic delayed reciprocity of patrilateral alliance could be plotted onto a diagram of symmetric alliance; but this would not justify the ascription of alternation as a general property of bilateral marriage. Another interpretation might be that a terminology of symmetric alliance can be socially manipulated, in the contraction of individual marriages, in accordance with a patrilateral preference; but this empirical possibility also does not make it true that bilateral marriage has the systematic characteristic of alternation.

Lévi-Strauss has thus published two kinds of statements, in his work on the elementary structures of kinship, about alternation. He has suggested that alternation may be a basic and immediate datum of mental and social reality; this proposition is not argued out, and the mode of relation is not analyzed. He has also made a number of assertions about the place of alternation in systems of affinal alliance; these propositions are groundless, irrelevant, or mistaken.

VII

Alternation occurs in such a variety of settings, and at such far-separated points in history and in global extent, as to make it reasonable to ask if it is not indeed a basic and immediate datum of mental and social reality.

Clearly, it is both "basic" and "immediate" in those systems or institutions which are necessarily defined, in part, by alternation itself. The constitution of Iatmül society, for instance, cannot be systematically represented without recourse to this mode of relation, and this feature is quite patent. More generally, wherever the relation is encountered in social forms it is a datum of as plain and unambiguous a character as a social fact can be. To qualify it as basic or immediate is to say no more, so far, about its incidence or function in social reality than we already know when we are tempted to describe it in these terms. If it is basic, it is so only in those settings in which we identify it; and if it is immediate, this means no more than that we can in fact identify it in the articulation of social forms. Certainly it is not basic in the sense of having a universal distribution, and it is not immediate in the logical sense of being inferable from a single premise without the introduction of a middle term. The definiteness or imprecision of the forms in which it is presented affects only the ease with which it can be determined, and this has nothing to do with its supposedly fundamental character as an aspect of social reality.

In any event, the assumption that alternation is basic does not in the least imply that it need not be explained. The concept of opposition, by comparison, certainly has good claim to be considered basic to collective representations, yet this very character has proved to call for deeper and deeper attempts at explanation (cf. Needham 1973a: xviii–xxxv; 1980a, chap. 2); it can be taken as a starting point only if the concept is employed so loosely that we cannot know where we are starting from. The case with alternation is far worse, since it has not yet received the comparative and formal treatment which has permitted some qualified reliance on the relation of opposition. And if alternation is to be accepted as a basic datum of "mental reality," then it is imperative to explain the deep grounds on which it should be accorded that ultimate status.

The problem, when rephrased in a more abstract way, is whether alternation is a principle. What is to be determined, therefore, is strictly speaking whether this mode of relation is primary (Lat. *princeps*, first in place or time)—that is, not derivable from any other relations or factors.

We can get an idea of what to look for by a comparison with the concept of duality. Kant took it as a fact that any *a priori* division of concepts must be by dichotomy (1787: 110), and this proposition can easily be conceded. Duality is indeed a principle, and dualism is accordingly the simplest conceivable form of classification. Is there any sense in which alternation can likewise be ascribed an *a priori* character?

The first step is to compose an adequate formal account of what is entailed in alternation; that is, to analyze the concept into those components which are necessary and sufficient to provide an integral formulation of the mode of relation. These components can be abstracted as:

1. events (e)
2. time scale (t)
3. sequence (s)
4. duality (1/2)
5. categories (a, b)
6. prescription ($e_n = e_{n+2}$)

The postulation of these components calls for only brief justification. It is self-evident that in registering alternation we are dealing with events (1), and thereby with a time scale (2); the events occur one after another, thus in sequence (3); by definition, any event can be in only one or another of two conditions (4); the contrastive names of these conditions supply the categories (5) which describe the process; and the regulation of the process is secured by the prescription (6). The interconnections of these components can be depicted as in figure 8.

Taken as it stands, this is a complex set of components, not a simple expression of the operation of a principle; at least, it is not an expression so simple as to be appropriate to an elementary principle.

We can however render the account less complex by canceling

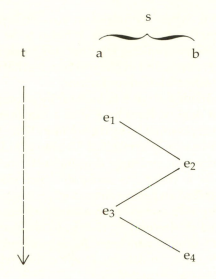

FIGURE 8.   *Interconnection of components of alternation.*

out those components which are premises of any form of social action and not peculiar to alternation; these are events (1), time scale (2), and categories (5). Perhaps in addition duality (4) can be accepted as an elementary proclivity of thought, as also it can be argued to be a logical necessity (cf. Blanché 1966: 15). There will then remain to be analyzed the notion of sequence (3), and crucially the form of the prescription (6) which governs the mode of sequence constituting alternation. This outcome poses a formal difficulty, in that the notion of a sequence is logically prior to that of alternation; so if alternation is a derivative or mode of sequence, it is in this respect not a principle.

There is a possible reinterpretation of the matter. Given that the operation of discerning (or conceiving, or constructing) a sequence is fundamental, then there are two elementary forms of sequence: successive and periodic. The former is continuous, the latter is interrupted. In the field of phenomena, such as social facts, it could however be said that any sequence is periodic. This is true in that something necessarily intervenes between repetitions of a kind of event; otherwise they could not be discriminated as events. But this consideration applies to both forms of sequence, and cannot distinguish between them. Also, the periodic sequence $e_1, -, e_3, -, e_5, \ldots$ is a series which, read from one term to the next, constitutes succession of a kind; the interstices correspond to intervening terms (events) of the same sort, but in a different context (another agent or location, etc.)—that is, they fall under the opposite category, either a or b. In these respects, alternation might be assimilated to an elementary form of sequence; and under this aspect it could be regarded as just as fundamental, or just as much a principle, as is the concept of a sequence. But of course this interpretation leaves out of account the prescription (6) which makes alternation into a distinct mode of sequence, and it is this distinctiveness that has to be accounted for.

Let us look at the negative aspect of the prescription. The converse of the formula $e_n = e_{n+2}$ is $e_n \neq e_{n+1}$. This negative formula ensures alternation only on the premise of duality (4); compare $e_n = e_{n+3}$, which obeys the prohibition but does not procure an alternation. With the premise of duality, however, the negative formula has a significant aspect: it is inconsistent with unison, as would be exemplified by figure 9. For unison is consistent with the prescription in the sequence e_1, e_3, e_5 when this is taken together with the parallel sequence e_2, e_4, e_6; that is, all the terms in a sequence are a, or else they are all b. What is essential to alternation, however, is the regular intercalation of events of the alternative category, and the grounds of this periodicity are still obscure.

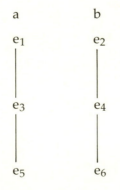

FIGURE 9.   *Sequences in unison.*

In the face of this persistent difficulty—and since in any event we should turn the matter deliberately, so as to display it under varied aspects—let us conclude this section by presenting the relation in another form, as in figure 10. This shows the same set of interconnections, of course, as in figure 8 above. The difference is that the events are represented as a single sequence, as in real social time, diacritically marked as belonging to category *a* or to category *b* in alternation. This form of diagram may appear more nearly isomorphic with the alternation of sections in a Kariera local line, or with the alternation of affinal lines in the Aranda or the Iatmül system, and from some point of view this could seem an advantage. However, the correspondence really depends more on certain conventions that are common to the diagrams of these systems; to begin with, a time scale represented as a vertical vector directed from top to bottom. Essentially, a representation that is both economical and adequate has been at hand from the beginning of section IV: namely, in the equation that formulates the prescription. The point, then, of an additional diagram is not to secure a closer approximation to the form of the relation, but to prevent the imagination from becoming fixated on a single representation as though it were the one correct and concrete form of the relation itself.

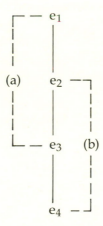

FIGURE 10. *Alternation as direct timelike sequence qualified by category.*

In fact, numerous other varieties of diagrams can be imagined, especially for the prescriptive systems, and there could
be advantages in constructing them and perhaps also in
switching methodically from one to another in order to procure changes of aspect under which the problem might be
viewed. But even this procedure incurs a risk, for when operat-
ing with diagrams there is a danger that we shall in some way
attribute the form of some such representation to the social
facts under study, or even to the consciousness of those whose
lives are in part ordered by what the diagram depicts.

As we approach the end of the present section, therefore, we
may profit from the explicit reminder that the object of our investigation is an abstraction, not a concrete form of any kind.
The concept of alternation can properly be said to have a structure, and one that is resistant to assimilation or reduction. It is
the constituent relations of this formal construct that need to be
accounted for if the grounds of alternation are to be uncovered.

VIII

In the case of certain prescriptive systems, it may
be possible to account for alternation, to a certain extent, by reference to standard features of such systems and to the consequences of introducing a change in the mode of contracting alliances.

Thus the Aranda and the Dieri systems can be derived hypothetically from a two-line system. Korn has singled out a principle, in alternating systems, according to which a line cannot
be affinally related to the same line in two consecutive genealogical positions (1973: 121). This negative factor in established
systems can also be recast as a determinant in bringing about a
change from a two-line to a four-line system. It is not significant, with regard to alternation, that the Aranda system is
framed by patrilines and the Dieri system by matrilines, for
this difference too could be accounted for by a hypothetical reconstruction of their evolution. The essential is the change
from two lines to four lines. Each of the resultant four lines
must have systematic functions, implicating the three other
lines, for otherwise the new arrangement would be equivalent

to nothing more than two adjoined and concomitant two-line systems. If all four lines are to be systematically interrelated, there automatically comes into play the alternation that is in fact characteristic of such systems as the Aranda and the Dieri.

But this still leaves undiscovered the cause of the change from two to four lines, for this is thereby the cause of the alternation also. As Korn has explained, a prohibition on any line's being affinally related to the same line in two consecutive genealogical levels can be ascribed to recognition of the mode of descent opposite to that of the terminological lines; but then this new affinal recognition needs in turn to be accounted for. In sum, it looks as though the systematic change in question is indeed systematic, and that alternation in this case is an incidental product of a number of determinants.

When we compare alternating prescriptive systems in areas between which there is no obvious or even likely historical connection, such as the Australian societies with the Iatmül, this incidental character of alternation is confirmed. In these cases the occurrence of alternation appears to be the result of convergence. The Australian systems are symmetric with four lines and the New Guinea system is asymmetric with five lines, but the structure of the alternating module (fig. 6) is the same. A formal comparative analysis simply determines this common structural component, but it does not explain its occurrence. Such an explanation is all the less conceivable, moreover, when we take into account the many other manifestations of alternation in Iatmül culture. The alternating division of clans into *mbapma* may have a functional connection with the regulation of marriage, but this cannot, on the evidence, be claimed for the alternation of the initiatory system; and the same form of order in sibling rivalries, the inheritance of personal names, and the performance of flute music has nothing directly to do with the prescriptive system.

I have started with prescriptive systems at this point because their simplicity and clear structure give promise of a systematic explanation of the feature of alternation. But the splendid case of Iatmül society, precisely by virtue of the extent to which its institutions are framed by alternation, frustrates that promise and concentrates attention back on the relation itself. If alternation is an incidental product of descent and affinity, in the evolution of Australian systems and in the Iatmül system, its determinants are likely to be various. Are we then to think that there are different grounds for each occurrence of an alternating institution, or is each instance the expression of some constant determining factor? In the case of the Iatmül, must we concede a different reason for alternation in quarrels and in the comity of flute-playing, or should we instead see each instance of alternation in this culture as an expression of an intrinsic and favored mode of relation?

The Iatmül case might be discounted, by the argument that these instances are not culturally differentiated; once a society has the idea of alternation, it might be contended, there is no need to account separately for each of the expressions of this relation. The question then is: How did the Iatmül get the idea in the first place?

Well, we do not know that they have the idea at all. All we know is that a number of their institutions are framed by alternation; we have no evidence that they contemplate this mode of relation in abstraction, or that they assimilate one institution to another by reference to alternation. What we are faced with, rather, is the issue whether any society anywhere has formed the idea of alternation in the organization of its corporate life, or whether contingent factors have conduced to this relation in one society after another. But, in the latter event, if alternation is an incidental product of convergence, why in so many societies with such different constitutions and concerns should merely contingent factors have converged precisely in such a way as to result in this distinct mode of relation?

Against this background, a more radical explanation appears instead to be considerably more likely: namely, that the distribution of alternating institutions, and the disparity of their settings (friezes, prescriptive systems, names, flutes, etc.), should be accepted as arguments for an intrinsic character in alternation that is independent of the particular kinds and forms of social facts in which the relation is expressed.

The possibility that alternation reflects a mental proclivity in the organization of experience may be strengthened by reconsidering it as a form of dyadic periodicity. Numerically, the next more complex form is that of triadic periodicity (e.g., 1, 2, 3, 1, 2, 3, . . .). Why is this not more common, or as common as alternation? Actually, it seems to be pretty uncommon, or at any rate not to have attracted the attention of ethnographers and comparativists. In the field of age-group organizations, which is very extensive, Frank Stewart remarks that "only deviant . . . systems have more than two streams" (1977: 127) such as are produced by alternate linking (cf. sec. V above). It might then be, he continues, that "a theory that accounted for the prevalence of alternate linking by pointing to the possibility of two streams would also have to explain why the possibility of having three or more streams is never (or hardly ever) exploited in age-set systems."

The possibility to which Stewart adverts is that alternate linking conveniently divides a succession of age-groups into two parties ("streams"), but this construction placed on the ethnographic facts merely restates the fact that this is what alternate linking does. What has to be explained is why exactly two streams should be desired, and not three. If two are common and have a global distribution, whereas three are apparently never found, this makes a case for some general determinant of dyadic periodicity. A search for this determinant, however, would divert us into a resumed investigation of duality, a matter that is notoriously obscure despite its remarkable prominence in collective representations. Duality is a necessary condition of alternation (cf. sec. VII above), and conversely, in the case of streaming, alternation produces a dual division of age-groups; but these facts do not explicate the mode of relation that is our concern, and a renewed concentration on duality would be unlikely to do so. What the facts in question may be taken to do is to confirm the supposition that a mental proclivity is at work—even if at one remove, as it were—behind the global recourse to dyadic periodicity that constitutes alternation.

It may be wondered, in this case, why such a proclivity should not be manifested even more commonly than alternation, in fact, is.

IX

A first response to this question, which has been subjacent to our entire investigation, is that we can make no prior estimation of what frequency would be proportionate to any inclination of the kind.

The only evidence we possess is the testimony of ethnographic reports, historical narratives, works of art, etc.; and it is the prominence of a certain feature among these social facts that leads to the idea of a proclivity that is responsible for its global and recurrent incidence. What is to count as "prominent," moreover, cannot be exactly stipulated, as by a statistical computation; and what is to count as a "global" distribution depends not only on a certain dispersal of the feature around the world but also on an assessment of the likelihood of diffusion (cf. Needham 1980a: 32–33).

Nevertheless, a certain pattern of social facts can well create the reasonable conviction that the feature in question should be traced to an origin that is intrinsically independent of institutions and the vagaries of historical happenings. Also, an argument by elimination is nonetheless a cogent form of proof in some circumstances; and when the explanation of the given feature can be had in no other way, there is no formal objection to the inference that the feature is a more or less spontaneous expression of a natural proclivity of thought or imagination.

We may compare the concept of a logical constant, such as a particular mode of syllogism; it does not appear in every rational proposition, and we do not expect it to do so, yet its minor incidence does not impair its character as a standard form of thought. Or we may compare the concept of an iconic archetype, such as the half-man (Needham 1980a, chap. 1) or the vagina dentata; it does not appear regularly and universally in all cultural contexts, yet its sporadic occurrence does not reduce its proper significance as a standard form of imagination.

So also, the relative infrequency with which alternation is resorted to in collective representations does not argue against its fundamental character. A mental proclivity, even a necessary one, is not necessarily discernible in all kinds of social facts whatever; just as Iatmül culture seems to favor alternation, so we can imagine a culture which forbade it. If alternation is a basic concept, in that it is the product of a natural inclination of the mind, still there are no grounds on which to contend that it ought to be more prevalent than it is.

A more serious doubt is of a formal kind. Alternation has

appeared to be not primary but derivative; that is, it is premised on other concepts. This has to be conceded, but there are at
least two defenses of the hypothesis that it is a basic concept in
the required sense.

The first defense is that the formulation of any principle perhaps necessarily resorts to the positing of things other than the
principle itself. Thus the operation of dichotomy as an *a priori*
act of the mind presupposes action, possibly an agent, an object susceptible to this action, ideas of one and two, perhaps
whole and part. The second defense is that a basic concept need
not be a principle; that is, it need not be an irreducible formal
relation (cf. symmetry, transitivity), but it may be a complex
notion. That such a concept may be analyzable into a set of
components, some of which may themselves be elementary
concepts, says nothing against its discrete and autonomous
character.

Compare, in the sphere of imagination, the synthetic image
of the witch (Needham 1978, chap. 2). This representation can
be analyzed into a number of primary factors of experience, but
the feasibility of doing this does not at all impair the characteristic form of the image or reduce the significance of its global
distribution. Likewise, the concept of alternation is certainly
synthetic, but this is not a formal ground on which to deny it a
fundamental character. We are misled in this matter, I think, by
an uncritical attachment to the criterion of simplicity. By this I
mean that although a principle of the kind we have been seeking must indeed be irreducible, and will be simple in this sense,
there is no reason that a basic concept shall be a principle and
thus be simple in the same way. Also, the expression of even a
simple concept may well need to be complex, and the combina-
torial formulation of a concept is easily consistent with a fundamental character. The concept of alternation, therefore, need
not be primary, and the formal expression of its constitution
need not be simple.

In the end, nevertheless, it is possible to remain subject to a
dual unease. On the one hand, there is our inability to account
formally for alternation; on the other, we should beware of con-
cluding too readily that the concept corresponds to a proclivity
for this form of order.

Methodologically speaking, however, there is no need to subscribe to either of the theoretical aims implied by these alternatives. For purposes of comparison and analysis it is enough to recognize the recurrence of alternation, together with the variegation of forms in which it can be manifested, among the modes of relation that frame collective representations. It would of course be more than satisfying if we could account for the concept on formal grounds, and thereby demonstrate that it responded to some cognitive necessity; and it would naturally be more convincing if we possessed unequivocal criteria by which to prove empirically that it was the expression, through the media of social facts, of an innate mental proclivity. Yet even if for the present we cannot attain that pitch of cogency in making the case for alternation, still we can now more readily accord it the dual interest that we began by noticing.

Alternation is indeed a basic concept in the conjoined senses stipulated (sec. I above): it has an immediate practical utility in the structural analysis of social facts, and it appears to be an index to a common property of the human mind.

7

Wittgenstein's Arrows

I

Among the modest technical resources of comparativism, a common aid to analysis is the use of diagrams: sometimes in order to sum up the constants in a verbal description, sometimes to delineate a structure, and for other purposes as well. One of the most frequently employed graphic devices is the arrow.

This directional indicator is particularly useful in representing transactions, such as the regular prestations in prescriptive systems. It looks very simple ($\rightarrow$) and its meaning would appear to be quite plain. There are various ways of drawing the arrow—with the fletching (⇒), for example, or without—but so long as it has a head and a tail it looks to be perfectly straightforward.

The arrow is not a natural sign, however, but conventional; the meaning ascribed to it has to be learned, and some knowledge of airborne missiles is needed in order to grasp its appositeness to that meaning.

To allude to the meaning of "arrow" leads us to the word itself and its derivation. It comes from late OE *ar(e)we*, the earlier form of which was *earh*; the Germanic base was **arχw-*, from IE **arkw-*, from which there also developed Lat. *arcus*, a bow (Onions 1966: 52, s.v.). The *Oxford English Dictionary* registers an argument that the Germanic word probably connoted "the thing belonging to the bow" (cf. Malay *anak panah*, arrow; lit. child [sc. complement] of the bow). The Germanic word for the bow itself was possibly connected with the names of certain trees or woods, though this is not certain. As for "arrow," however, we note already that the word comes not from the missile itself, but from the weapon from which it was discharged.

The derivations of words equivalent to "arrow" in other languages are not so unexpected, but they nonetheless convey different formative ideas. The German *Pfeil* comes from Lat. *pīlum*, javelin (Grebe 1963: 54, s.v.), thus making no reference to the bow and assimilating the arrow to a different missile. The French *flêche* probably comes from Frankish **fliukka*, arrow, missile; and this word in turn is connected to Middle Dutch *vlieke*, quill, pinion, also a weapon that is thrown or launched (cf. Verdam 1956: 720, s.v. *vlieke*). The noun **fliukka* probably derives from Germ. **fleukkon*, from the verb **fleugnôn*, to fly (*Trésor* . . . 8 [1980]: 966, s.v. *flêche*). The emphasis in this case is on the idea of flight. The Avestan *tiγri*, arrow, meant "sharp-pointed." In Sanskrit there are two words: *śaru*, a missile weapon, often an arrow, sometimes a spear; and *bāṇa*, arrow, equivalent to *vaṇa*, a reed. Since *śaru* is related to *śara*, also a reed, the central image in each case is that of the plant from whose stems arrows were made (Buck 1949, art. 20.25). As a final example, the Chinese character *chien*, arrow, is written with the radical 118, *chu*, bamboo; so that in this case also, the stress is apparently placed on the material that provides the shaft of the arrow.

To judge by these etymological indications, there is no single or definitive attribute of an arrow that constitutes a common implicit meaning, even in linguistic traditions that were equally familiar with the form and capabilities of the missile. Similarly, when the arrow is taken symbolically, it can of course be given quite contrasted meanings. I do not refer to the obvious and expectable fact that different traditions may confer various mythical and figurative connotations on the missile; for example, that it is made to stand for the sudden affliction of love (Kāma, Eros) or for a rapid access of understanding (*Mundaka Upaniṣad*). What I have in view is the possibility that the arrow, by virtue of some property or resemblance, may evoke a subconscious response of a more fundamental kind; for instance, that its form or activity might make it seem appropriate to express a certain significance that was not due to its elaboration in a particular civilization. Thus its rigid length and power of penetration could ascribe to it a phallic connotation; and this is actually expressed in the now obsolete usage (testified to in the fourteenth century) whereby "arrow" could mean "penis" (*O.E.D.*, s.v., II. 5).

It is the more decisive therefore that there is textual evidence, from traditional India, that "the vulva, the arrow, and the number 5 are considered equivalent symbols" (Daniélou 1964: 265); e.g., Pārvatī, the beloved of Śiva, "is represented by the number 5 (i.e., has the form of an arrow)" and once "took the shape of a vulva" to catch hold of a phallus of fire that was burning up the world.

These considerations should be enough, then, to avert the presumption that the arrow has intrinsically some steady inner significance making it apt to stand for directed relationships in anthropological diagrams.

II

Wittgenstein, in *The Brown Book*, pays long attention to the concept of similarity. As one example, he examines the case of someone who contends that there are correspondences between light or dark colors and particular vowels. He concludes that in certain circumstances we shall be inclined to talk of different relations, in certain others to talk of the same relation. "One might say, 'It depends how one compares them'" (Wittgenstein 1958: 138–40).

Then he asks the provocative question (p. 140):

Should we say that the arrows → and ← point in the same direction or in different directions?

His response is that at first sight we might be inclined to say that of course they point in different directions. But, he goes on, we can look at it in the following way.

If I look into a mirror and see the reflection of my face, this can be taken as a criterion for seeing my own head. If I happened to see the back of a head, I might say that this could not be my own head I am seeing but a head looking in the opposite direction. "Now this could lead me on to say that an arrow and the reflection of an arrow in a glass have the same direction when they point towards each other, and opposite directions when the head of one points to the tail end of the other."

Wittgenstein imagines another case. A man has been taught the ordinary use of the words "the same" in such phrases as "the same color," "the same shape," "the same length"; and he has also been taught the use of the words "to point to" in such contexts as "the arrow points to the tree." Then he is shown two arrows facing each other, and also two arrows one following the other, and he is asked to which of these two cases he would apply the phrase "the arrows point the same way." Wittgenstein clinches this example by asking: "Isn't it easy to imagine that if certain applications were uppermost in his mind, he would be inclined to say that the arrows → ← point 'the same way'?" (p. 140).

In the exercise that follows, I want to look further at some of the different aspects under which the pointing of an arrow can be interpreted, and to remark incidentally some implications of using arrows to represent social facts.

III

Let us start with the simplest case, and with it establish our descriptive conventions.

Here is an arrow such as might be used in a diagram:

$$(1) \quad \rightarrow$$

The end equipped with barbs is its head, the other end is the tail; the ends define its axis. The arrow is directed in relation to coordinates that we need to fix; let us say, as in heraldry or in stage directions, that it goes from proper (or stage) right to proper left.

It may not simply point; the direction may be intended to state that something is transferred in that direction or that something less material is acted out towards the head. While this particular arrow, directed as it is, will be taken as our para-

digm, it should be understood that neither its form nor its attitude is of any analytical importance. The arrow can be of any
shape or any degree of Steinbergian elaboration, just so long as
it has only one end that is recognizable as the head. The head
can be to the right or to the left, or the arrow can be vertical or
else inclined at any other angle. When two arrows are compared, they can be in any attitude and in any positions relative
to each other, so long as they lie along the same axis or on parallel axes.

Wittgenstein's first example takes its place here in the form
of the following two arrows:

$$(2) \quad \rightarrow \rightarrow$$

At first sight we should say that they point in the same direction. Wittgenstein makes us imagine the second arrow as
reflected in a mirror, in which event it would be reckoned as
pointing in the opposite direction to the first. How legitimate
an example is this? Is it not a trick? After all, an arrow head-on
to a mirror is reversed in its reflection, and we do not normally
judge the directions of things by how they would appear if they
happened to be seen reflected in mirrors.

Yet it is not a trick of language that permits us to say that the
arrows in no. 2 above would be pointing in opposite directions
if the second were the reflected image of the first. We might say
that what Wittgenstein has done is to introduce a nonformal
factor which allows us to grasp certain coordinates by which
we are enabled to determine in what sense the arrows can be
said to point in the same direction or not. That this contextual
factor is a mirror does not falsify the coordinates: it provides a
valid and necessary ground for an intelligible employment of
the words "same direction" and "opposite direction."

At the limit, it could be said that there is no *a priori* significance in the unqualified use of the words "same" and "opposite." In the case of the arrows in no. 2, the answer does indeed
depend on how one compares them. A comparison of one arrow with the other depends on coordinates which are independent of the arrows themselves. A two-dimensional plane provides one set of coordinates according to which the second
arrow is formally a translation of the first and must therefore
point in the same direction. A mirror introduces a perpendicu-

lar axis between the arrows and calls for a rotation of the first arrow; if it does not rotate, it points in the opposite direction to the proper consequence of rotation—and thus opposite to the direction of the original arrow.

There is in fact a picture by Magritte ("La Reproduction interdite") which makes these points. A man is looking into a mirror and sees the back of what is plainly his own head; the viewer can see this too. In one sense both heads are looking in the same direction, like two men standing one behind the other in a queue; in another sense, once a plane of rotation has been introduced by a mirror, they are looking in opposite directions. The idea of a direction depends on coordinates, and until these are circumstantially determined the expressions "same direction" and "opposite direction" cannot be positively interpreted. Another way to put the issue is this: Without a setting, it is only formally true that the arrows in no. 2 point in the same direction; but the interpretation of phenomena, whether social or other facts, is not a formal discipline. If the figure is taken to represent an empirical state of affairs, and thereby to have a content, it can be seen under different aspects according to the properties of whatever is represented.

Wittgenstein's second example provides the following case:

$$(3) \quad \rightarrow \leftarrow$$

Considered formally, the arrows evidently point in opposite directions: one from right to left, the other from left to right. But if we supply a content, in the form of some point of reference located (like the tree in Wittgenstein's example) between the heads of the arrows, then it can be said with equal validity that the arrows point the same way. In this case there is no seeming trickery such as the reflecting plane of a mirror: the changes of aspect depend simply on the stipulation of a point of reference. Adopt two such points, one behind the nock of each of the arrows, and with only this attribution of a setting the arrows point again in opposite directions.

Now let us introduce the next formal manipulation, with this case:

$$(4) \quad \leftarrow \rightarrow$$

Taken as they are presented, the arrows point by definition in opposite directions; one goes (as defined by the position of the head) from left to right, the other goes from right to left. But imagine them as drawn on a vertical cylinder, with a point of reference (a target, as it were) on the other side and, for the sake of simplicity, located diametrically opposite the gap between the nocks of the arrows. In this event it can correctly be said that the arrows point in the same direction: each is directed towards the same point on the other side of the cylinder.

If it should be objected that this is a contrived proof, the first answer is that it is a cogent demonstration all the same. What we have done is simply to provide two items of content: a continuous curved surface instead of a plane, and a point of reference on this surface. Since the arrows then point towards the same target, they point in the same direction. Under a formal aspect they point in opposite directions; by a particular substantive interpretation they point in the same direction.

It might be objected that a hypothetical construction such as an imaginary cylinder implies nothing of real importance as far as concrete cases are concerned, but this would be incorrect. In the first place, the application of the construct cannot be determined in advance of the phenomena to which it may apply; these are in principle unpredictable, and so therefore is the potential utility of the construct. Moreover, there exist social phenomena which are represented conventionally in a way that corresponds directly to the present example. The forms of social classification known as asymmetric prescriptive alliance are directional and closed, so that in certain respects example 4 accords very well with their distinctive features.

With regard to the continuous curved surface, there is a categorical closure in systems of this kind such that Bateson was led to propose a technique of circular diagrams; the terms and their connectives were to be "rolled so as to form a cylinder around something hard, such as a jam jar" (Korn 1973: 97, n. 14; cf. chap. 6 above, fig. 5). As for the directions of the arrows, it is a standard feature of asymmetric systems that their symbolic prestations are divided into two classes, often named, such as "sword"/"cloth," "beasts and ear pendants"/"cloths and pigs," "food"/"valuables" (Korn 1973: 100–03), etc. These have been described respectively as "masculine" goods in opposition to "feminine" goods. Masculine goods circulate in one direction, and feminine goods circulate in the opposite direction, just as in example 4. If a particular recipient of both masculine and feminine goods is fixed upon (cf. the point of reference on the other side of the cylinder), then, viewed from the point of origin (cf. the gap between the nocks of the arrows), the prestations are transmitted in the same direction, namely towards that recipient.

Admitted, in such cases a formal construct is being related to a formal representation of social facts, so it is not to be claimed that example 4 corresponds directly to "social reality"; but it does correspond nicely with a proved analytical representation of certain social phenomena. An unexpectable interpretation of example 4, namely, that the arrows point in the same direction, has therefore a direct application, and possibly some utility, in the study of social facts.

Now let us revert to a single arrow and consider this example:

(5) ←

By definition, this has only one direction; the head is to proper right, and that is the direction in which it points. Taken in isolation, as it is presented, it cannot be said to go in the same direction as anything else, nor in the opposite direction. But put a frame about it as follows:

(6) [←

Imagine that the frame is a quiver. The head of the arrow remains to the right, but if a movement is to be imparted to the arrow it cannot travel to the right (that is, through the bottom of the quiver). The only direction in which it could move is to the left, out of the open end, and this of course is the way an arrow is actually withdrawn from a quiver. So if we ascribe to the formal construct the attribute of potential movement, the direction taken by the arrow is opposite to that in which the head points.

The example can also be interpreted statically, yet if anything more convincingly. Imagine this time that the frame represents the ground, and that the arrow has been thrust point-forward into the ground, leaving the shaft standing vertically above the surface. (It will be recalled that the attitude of the arrow in the diagram is of no importance.) Here we have what in surveying is actually called an "arrow"; namely, a straight stick shod with iron (originally a real arrow), or an iron pin, used to stick in the ground at the end of each chain. It does not merely secure the chain, but it can serve as a marker; and we can correctly say that under this aspect the arrow points (up in the air) in the opposite direction to that in which the head points.

In principle the converse obtains, as in this case:

$$(7)\ \boxed{\rightarrow}$$

Formally, the arrow points to the left, out of the frame. There might also be a real setting in which the arrow could be seen as pointing nonetheless to the right, tail-first; but I am not acquainted with an actual usage, involving the manipulation of an arrow, that would be interpreted in this sense. It would be surprising, however, if there were no such case to be found in some society. We can easily imagine, at any rate, that a people accustomed to the constant carriage of bows and arrows might use an arrow for the purpose of directing attention to things, but that it would be bad form to point with the barbed head.

Before we leave these examples of arrows and the directions in which they can be said to point, it is worth mentioning graphic arrows that do not point at all. Here is one:

$$(8)\ \Downarrow$$

This is a schematic representation of what in heraldry is called a pheon; that is, a broad-arrow used as a badge or as a component in a coat of arms. Usually it is point-down, but occasionally the blazon may stipulate that it be point-up. It stands for whatever significance may have accreted to it historically in connection with a particular coat of arms. It does not point towards anything, and it is not understood as conveying a movement in either direction: it simply presents itself as an emblem.

The broad-arrow has also been used in the following attitude:

$$(9)\ \Uparrow$$

"In the King's Forests they set the Figure of a broad Arrow upon Trees that are to be cut down"; hence Sir Thomas Browne could write in 1690 of Death's broad arrow (Browne 1968: 94). The emblem used to be employed by the British Board of Ordnance and was placed on government property; and it was also displayed on clothing issued to convicts. Here the arrow is used as a proprietory sign; although it is described as pointing upwards, it does not actually point to anything and it has nothing to do with direction.

These broad-arrow examples reinforce the argument that an arrow does not possess an intrinsic significance, even in the eyes of those who are familiar with the object from which the graphic device is taken. If it is to convey a meaning, that meaning has to be explicitly assigned to it; and the meaning may then be quite different from that which at first seemed appropriate to its form.

IV

The cases briefly considered in the preceding section constitute an exercise in changes of aspect.

Taken simply as an exercise, the survey of examples may have a useful effect in conducing to a relaxation of the mind in the face of diagrams; the imagination can be given more play, while at the same time the analytical intelligence is pressed towards more rigorous stipulations in the construction and interpretation of diagrams. But there is also a practical side to the matter, for anthropological diagrams are supposed to be useful in the representation of social facts; and the more ambiguity we can discover in a graphic device, the more precise we can be in exploiting that utility.

At one place above (example 4), I have adduced some characteristic features of a certain type of social system, mainly for the purpose of demonstrating that even an unlikely interpretation of a diagram can have a real application. I do not claim that every interpretation reveals an aspect that has a real counterpart among social facts, but only that it may do so; also that until we have methodically turned a diagram about in our minds, so as to reveal as many aspects as possible, we shall not know what its range of application may be. We need in this way to "look through" our technical apparatus, and not treat it as though it were a concrete replica of social facts. A diagram, whether an arrow or a genealogical scheme or a matrix, does not depict a structure: it permits a structure to be conceived (Needham 1980b: 78). How we conceive it depends in part on our imaginative agility, as also on our criteria of similarity and correspondence. From these considerations derives the analytical resourcefulness of Wittgenstein's arrows.

Against the chance that the present exercise with arrows may nevertheless be taken as largely a formalistic diversion, let me conclude with a factual case from social anthropology.

It might be held that in practice there is no problem about the employment of arrows in diagrams of social systems. Imagination can admittedly contrive all kinds of hypothetical interpretations, but when it comes to what people on the ground actually do, and the immediate description of what they do, the task is simple enough. If we see a diagram such as A → B, there is no real doubt of its meaning: the arrow means that A gives something to B, or performs some other action directed towards B; perhaps, more generally, that A is active and B passive. At any rate, whatever the social content of the state of affairs represented by the diagram, there is no ambiguity in the direction of the transaction indicated by the arrow: it goes *from* A *to* B.

Lévi-Strauss's monograph on the elementary structures of kinship (1969) is in effect all about the exchange of women, and the constant motif of his argument has to do with men giving women to other men. This kind of transaction is purported to be the agent in man's transition from nature to culture; it is the origin and precondition of civilization. Among the contemporary systems of affinal exchange that Lévi-Strauss surveys, some are symmetric in their organization and rely on a direct exchange of women; others are asymmetric and rely on an "indirect exchange" whereby women are transmitted in only one direction. Examples of asymmetric systems include certain societies on the Indo-Burma border, such as the Chawte and the Tarau, both of which are held by Lévi-Strauss to exhibit the characteristic "marriage cycles" of indirect exchange. The Chawte system, he states, consists of "a quaternary cycle within which there are two ternary cycles" (p. 271). We can ignore this precise-sounding characterization, since a correct analysis of Chawte ethnography reveals that the evidence is too confused and contradictory to be of any secure use in comparative studies or in theoretical analysis (Needham 1960b: 251). But Lévi-Strauss's treatment of this case does include one remarkable feature (n. 11 in my analysis) involving the use of arrows in his diagram of the cycles, which is carried over into the analysis of the Tarau system as well.

The Tarau, writes Lévi-Strauss, are divided into four lineages united by a simple relationship of generalized (= indirect) exchange: a Pachana man marries a Tlangsha woman; a Tlangsha man marries a Thimasha woman; a Thimasha man marries a Khulpu woman; a Khulpu man marries a Pachana woman (pp. 271–72). If we read this description in reverse, in order to grasp the system better, not as men *taking* women in marriage but as groups of men *giving* women to other men, the relationships read as follows:

Pachana   *gives to* Khulpu
Khulpu   *gives to* Thimasha
Thimasha *gives to* Tlangsha
Tlangsha *gives to* Pachana

Now this is how Lévi-Strauss (1969: 270, fig. 47) constructs his diagram of these alliances:

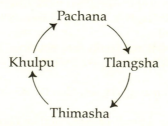

That is, the direction of the transaction whereby A *gives* women *to* B is represented by an arrow pointing this way:

A ← B

Bibliography

Adanson, Michel
  1763   Familles des plantes. 2 vols. Paris.
Adriani, N., and Alb. C. Kruyt
  1950–51   De Bare'e-sprekende Toradjas van Midden-Celebes. 2d
            ed., rev. by Alb. C. Kruyt. 3 vols. Amsterdam: Noord-Hol-
            landsche Uitgevers Maatschappij.
Agassiz, Louis
  1859   An Essay on Classification. London.
Apte, Vaman Shivram
  1914   The Student's English-Sanskrit Dictionary. Bombay:
            Sagoon.
Ayer, A. J.
  1954   Philosophical Essays. London: Macmillan.
Babcock, Barbara, ed.
  1978   The Reversible World: Symbolic Inversion in Art and So-
            ciety. Ithaca, N.Y.: Cornell University Press.
Bambrough, Renford
  1961   Universals and family resemblances. Proceedings of the
            Aristotelian Society 60: 207–22.
Barden, Garrett
  1976   Discriminating classes. Man 11: 345–55.
Barnes, R. H.
  1974   Kédang: A Study of the Collective Thought of an Eastern
            Indonesian People. Foreword by Rodney Needham. Ox-
            ford: Clarendon Press.
Bateson, Gregory
  1935   Music in New Guinea. The Eagle: A Magazine supported

          by the Members of St. John's College, Cambridge, 48:
          158–70.
1936   Naven. Cambridge: At the University Press.
Beckner, Morton
1959   The Biological Way of Thought. New York: Columbia
          University Press.
Beidelman, T. O.
1963   Witchcraft in Ukaguru. In: J. Middleton and E. H. Winter,
          eds., Witchcraft and Sorcery in East Africa (London:
          Routledge & Kegan Paul): 58–98.
Bernardi, Bernardo
1959   The Mugwe, a Failing Prophet. London: Oxford Univer-
          sity Press, for the International African Institute.
Blanché, Robert
1962   Axiomatics. Trans. G. B. Keene. London: Routledge & Ke-
          gan Paul; New York: Dover Publications.
1966   Structures intellectuelles: essai sur l'organisation systé-
          matique des concepts. Paris: Vrin.
Borges, Jorge Luis
1965   Other Inquisitions, 1937–1952. Trans. Ruth L. C. Simms.
          New York: Simon & Schuster.
Browne, Thomas
1643   Religio Medici. London.
1658   Hydriotaphia: Urne-Buriall. London.
1968   Selected Writings. Ed. Geoffrey Keynes. London: Faber
          & Faber.
Buck, Carl Darling
1949   A Dictionary of Selected Synonyms in the Principal
          Indo-European Languages. Chicago: University of Chi-
          cago Press.
Campbell, K.
1965   Family resemblance predicates. American Philosophical
          Quarterly 2: 238–44.
Castaneda, Carlos
1968   The Teachings of Don Juan. Berkeley and Los Angeles:
          University of California Press.
Collingwood, R. G.
1939   An Autobiography. Oxford: Clarendon Press.
Crevel, René
1966   Le Clavecin de Diderot. (1st ed., 1932.) Paris: Pauvert.
Crocker, J. Christopher
1977   The mirrored self: identity and ritual inversion among
          the eastern Bororo. Ethnology 16: 129–45.

Crooke, W.
1894   Introduction to the Popular Religion of Northern India.
        Allahabad.
Daniélou, Alain
1964   Hindu Polytheism. (Bollingen Series no. 73.) New York:
        Pantheon Books.
DeMille, Richard
1978   Castaneda's Journey: The Power and the Allegory. Lon-
        don: Sphere Books.
Downs, R. E.
1956   The Religion of the Bare'e-speaking Toradja of Central
        Celebes. The Hague: Uitgeverij Excelsior.
Dumont, Louis
1966   Descent or Intermarriage? A Relational View of Austra-
        lian Section Systems. Southwestern Journal of Anthro-
        pology 22: 231–50.
Durkheim, Émile
1895   Les Règles de la méthode sociologique. Paris.
1901   Les Règles de la méthode sociologique, 2d ed. Paris.
1951   Jugements de valeur et jugements de réalité. (Orig. pub.,
        1911.) In: Durkheim, Sociologie et philosophie (Paris:
        Presses Universitaires de France): 117–41.
Durkheim, Émile, and Marcel Mauss
1903   De quelques formes primitives de classification. Année
        Sociologique 6: 1–72.
1963   Primitive Classification. Trans. and ed. with an Introduc-
        tion by Rodney Needham. Chicago: University of Chi-
        cago Press.
Elshout, J. M.
1926   De Kĕnja-Dajaks uit het Apo-Kajangebied: Bijdragen tot
        de Kennis van Centraal-Borneo. The Hague: Martinus
        Nijhoff.
Endicott, Kirk
1979   Batek Negrito Religion. Oxford: Clarendon Press.
Evans-Pritchard, E. E.
1965   The comparative method in social anthropology. In:
        Evans-Pritchard, The Position of Women in Primitive So-
        cieties, and Other Essays in Social Anthropology (Lon-
        don: Faber): 13–36.
Fortune, R. F.
1933   A note on some forms of kinship structure. Oceania 4: 1–9.
Fox, James J.
1973   On bad death and the left hand: a study of Rotinese sym-

bolic inversions. In: Rodney Needham, ed., Right & Left: 342–68.

Frege, Gottlob
1960   Translations from the Philosophical Writings of Gottlob Frege. Trans. and ed. Peter Geach and Max Black. 2d ed. Oxford: Basil Blackwell.

Fürer-Haimendorf, C. von
1939   The Naked Nagas. London: Methuen.

Furness, W. H.
1902   The Home-Life of Borneo Head-hunters. Philadelphia: Lippincott.

Gaidoz, H.
1893   Les Pieds ou les genoux à rebours. Mélusine 6: 172–75.

Gennep, A. van
1909   Les Rites de passage. Paris: Nourry.
1960   The Rites of Passage. Trans. Monika B. Vizedom and Gabrielle L. Caffee. London: Routledge & Kegan Paul.

Gilmour, J. S. L.
1951   The development of taxonomic theory since 1851. Nature 168: 400–02.

Grebe, Paul, ed.
1963   Etymologie: Herkunftswörterbuch der deutschen Sprache. (Der Grosse Duden 7.) Mannheim: Bibliographisches Institut.

Hallpike, C. R.
1979   The Foundations of Primitive Thought. Oxford: Clarendon Press.

Hampshire, Stuart
1959   Thought and Action. London: Chatto & Windus.

Hardeland, Aug.
1859   Dajacksch-Deutsches Wörterbuch. Amsterdam.

Harner, Michael J.
1973   The Jivaro: People of the Sacred Waterfalls. London: Robert Hale.

Hose, Charles, and William McDougall
1912   The Pagan Tribes of Borneo. 2 vols. London: Macmillan.

Hull, Eleanor
1904   The story of Deirdre, in its bearing on the social development of the folk-tale. Folk-lore 15: 24–39.

Hutton, J. H.
1938   A Primitive Philosophy of Life. (Frazer Lecture, 1938.) Oxford: Clarendon Press.

Izikowitz, K. G.
1941   Fastening the soul: some religious traits among the La-

met. Göteborgs Hogskolas Årskrift 47: 14.

1951   Lamet: Hill Peasants in French Indo-China. Göteborg: Etnografiska Museet.

Jevons, W. Stanley

1874   The Principles of Science: A Treatise on Logic and Scientific Method. 2 vols. London: Macmillan.

Kant, Immanuel

1787   Kritik der reinen Vernunft. 2d ed. Riga, Latvia.

Karlgren, Bernhard

1923   Analytic Dictionary of Chinese and Sino-Japanese. Paris: Geuthner.

Korn, Francis

1973   Elementary Structures Reconsidered: Lévi-Strauss on Kinship. London: Tavistock Publications; Berkeley and Los Angeles: University of California Press.

Kruyt, Alb. C.

1906   Het Animisme in den Indischen Archipel. The Hague: Martinus Nijhoff.

1914   Indonesians. In: J. Hastings, ed., Encyclopaedia of Religion and Ethics 7: 232–50. Edinburgh: T. & T. Clark.

1973   Right and left in central Celebes. (First pub., in Dutch, 1941.) In: Rodney Needham, ed., Right & Left: 74–91.

Kükenthal, W.

1896   Forschungsreise in den Molukken und in Borneo. Frankfurt-am-Main.

Lancre, Pierre de

1613   Tableau de l'inconstance des mauvais anges et démons. Paris.

Lévi-Strauss, C.

1969   The Elementary Structures of Kinship. Rev. ed., trans. from the French by J. H. Bell, J. R. von Sturmer, and Rodney Needham; ed. Rodney Needham. Boston: Beacon Press.

Lévy-Bruhl, L.

1949   Les Carnets de Lucien Lévy-Bruhl. Paris: Presses Universitaires de France.

Lichtenberg, Georg Christoph

1968   Schriften und Briefe, vol. 1: Sudelbücher. Ed. Wolfgang Promies. Munich: Carl Hanser.

1971   Schriften und Briefe, vol. 2: Materialhefte, Tagebücher. Ed. Wolfgang Promies. Munich: Carl Hanser.

Lockhart, W. R., and P. A. Hartman

1963   Formation of monothetic groups in quantitative bacterial taxonomy. Journal of Bacteriology 85: 68–77.

Lowie, Robert H.
1917   Culture and Ethnology. New York: Boni & Liveright.
Lukes, Steven
1973   Emile Durkheim: His Life and Work. London: Allen Lane
            The Penguin Press.
Mauss, Marcel
1925   Essai sur le don: forme et raison de l'échange dans les
            sociétés archaïques. Année Sociologique n.s. 1: 30–186.
Mayr, Ernest
1969   Principles of Systematic Zoology. New York: McGraw-
            Hill.
Middleton, John
1973   Some categories of dual classification among the
            Lugbara of Uganda. In: Rodney Needham, ed., Right &
            Left: 369–90.
Monier-Williams, Monier
1956   A Sanskrit–English Dictionary. Oxford: Clarendon
            Press.
Morris, Ivan
1979   The World of the Shining Prince: Court Life in Ancient
            Japan. Harmondsworth: Penguin.
Morris, Marshall
1981   Saying and Meaning in Puerto Rico. Oxford: Pergamon
            Press.
Murdock, George Peter
1949   Social Structure. New York: Macmillan.
Myerhoff, Barbara G.
1978   Return to Wirikuta: ritual reversal and symbolic continu-
            ity in the peyote hunt of the Huichol Indians. In: Barbara
            Babcock, ed., The Reversible World: 225–39.
Needham, Rodney
1954   The system of teknonyms and death-names of the Penan.
            Southwestern Journal of Anthropology 10: 416–31.
1956   Ethnographical notes on the Siwang of central Malaya.
            Journal of the Malayan Branch, Royal Asiatic Society, 29:
            49–69.
1958   Notes on Baram Malay. Journal of the Malayan Branch,
            Royal Asiatic Society, 31: 171–75.
1960a   The left hand of the Mugwe: an analytical note on the
            structure of Meru symbolism. Africa 30: 20–33. [Re-
            printed in Needham, ed., Right & Left (1973), chap. 7.]
1960b   Chawte social structure. American Anthropologist 62:
            236–53.
1962   Notes on comparative method and prescriptive alliance.

Bijdragen tot de Taal-, Land- en Volkenkunde 118: 160–82.

1966   Terminology and alliance, I: Garo, Manggarai. Sociolo-
            gus 16: 141–57.
1967   Terminology and alliance, II: Mapuche, conclusions. So-
            ciologus 17: 39–53.
1970a   Endeh II: test and confirmation. Bijdragen tot de Taal-,
            Land- en Volkenkunde 126: 246–58.
1970b   Introduction to: A. M. Hocart, Kings and Councillors, ed.
            Rodney Needham (Chicago: University of Chicago
            Press): xiii–xcix.
1971a   ed., Rethinking Kinship and Marriage. London: Tavistock
            Publications.
1971b   Remarks on the analysis of kinship and marriage. In:
            Needham, ed., Rethinking Kinship and Marriage: 1–34.
            [Reprinted, with additions, in Needham, Remarks and
            Inventions (1974a), chap. 1.]
1971c   Penan friendship-names. In: The Translation of Culture,
            ed. T. O. Beidelman (London: Tavistock Publications):
            203–30.
1972   Belief, Language, and Experience. Oxford: Basil Black-
            well; Chicago: University of Chicago Press.
1973a   ed., Right & Left: Essays on Dual Symbolic Classification.
            Chicago: University of Chicago Press.
1973b   Prescription. Oceania 43: 166–81.
1974a   Remarks and Inventions: Skeptical Essays about Kinship.
            London: Tavistock Publications; New York: Harper &
            Row.
1974b   The evolution of social classification: a commentary on
            the Warao case. Bijdragen tot de Taal-, Land- en
            Volkenkunde 130: 16–43.
1975   Polythetic classification. Man 10: 349–69.
1976   Skulls and Causality. Man 11: 71–88.
1978   Primordial Characters. Charlottesville: University Press
            of Virginia.
1979   Symbolic Classification. Santa Monica: Goodyear Pub-
            lishing. (Distributed by Random House, New York.)
1980a   Reconnaissances. Toronto: University of Toronto Press.
1980b   Diversity, structure, and aspect in Manggarai social clas-
            sification. In: R. Schefold et al., eds., Man, Meaning, and
            History (The Hague: Martinus Nijhoff): 53–81.
1980c   Principles and Variations in the Structure of Sumbanese
            Society. In: J. J. Fox, ed., The Flow of Life (Cambridge,
            Mass.: Harvard University Press), chap. 1 (pp. 21–47).

1981   Circumstantial Deliveries. Berkeley and Los Angeles: University of California Press.

Nietzsche, Friedrich
1886   Jenseits von Gut und Böse: Vorspiel einer Philosophie der Zukunft. Leipzig.

Ogden, C. K.
1932   Opposition: A Linguistic and Psychological Analysis. Cambridge: Orthological Institute.

Onions, C. T., ed.
1966   The Oxford Dictionary of English Etymology. Oxford: Clarendon Press.

Opie, Iona and Peter
1959   The Lore and Language of Schoolchildren. Oxford: Clarendon Press.

Peacock, James L.
1978   Symbolic reversal and social history: transvestites and clowns of Java. In: Barbara Babcock, ed., The Reversible World: 209–24.

Rhees, Rush
1970   Discussions of Wittgenstein. London: Routledge & Kegan Paul.

Rivière, Peter
1971   Marriage: a reassessment. In: Rodney Needham, ed., Rethinking Kinship and Marriage: 57–74.

Simpson, George Gaylord
1961   Principles of Animal Taxonomy. New York: Columbia University Press.

Smith, Jonathan Z.
1970   Birth upside down or right side up? History of Religions 9: 281–303.

Sneath, Peter H. A.
1962   The construction of taxonomic groups. In: G. C. Ainsworth and P. H. A. Sneath, eds., Microbial Classification (Cambridge: At the University Press): 289–332.

Sokal, Robert R., and Peter H. A. Sneath
1963   Principles of Numerical Taxonomy. San Francisco: W. H. Freeman.

Stewart, Dugald
1810   Philosophical Essays. Edinburgh.

Stewart, Frank Hamilton
1977   Fundamentals of Age-Group Systems. New York: Academic Press.

Toulmin, Stephen
1969   Ludwig Wittgenstein. Encounter 32: 58–71.

Tournier, Michel
1970   Le Roi des aulnes. Paris: Gallimard.
Trésor . . .
1971–   Trésor de la langue française. Paris: Éditions du Centre National de la Recherche Scientifique.
Verdam, J.
1956   Middelnederlandsch Handwoordenboek. 2d ed. The Hague: Martinus Nijhoff.
Vygotsky, L. S.
1962   Thought and Language. Ed. and trans. Eugenia Hanfmann and Gertrude Vakar. Cambridge, Mass.: M.I.T. Press.
Waismann, F.
1968   How I See Philosophy. Ed. R. Harré. London: Macmillan.
Weyl, Hermann
1952   Symmetry. Princeton, N.J.: Princeton University Press.
Wilken, G. A.
1912   De Verspreide Geschriften. Ed. F. D. E. van Ossenbruggen. 4 vols. Semarang, Surabaya, and The Hague: G. C. T. van Dorp.
Wisdom, John
1965   Paradox and Discovery. Oxford: Basil Blackwell.
Wittgenstein, Ludwig
1953   Philosophical Investigations. Trans. G. E. M. Anscombe. (3d ed., 1967.) Oxford: Basil Blackwell.
1958   The Blue and Brown Books. Oxford: Basil Blackwell.
1967a   Zettel. Trans. G. E. M. Anscombe, ed. G. E. M. Anscombe and G. H. von Wright. Oxford: Basil Blackwell.
1967b   Bemerkungen über Frazers The Golden Bough. Ed. Rush Rhees. Synthese 17: 233–53. [Reprinted with corrections and English translation by A. C. Miles in: Remarks on Frazer's Golden Bough (Retford, Nottinghamshire: Brynmill Press, 1979).]

Index

Accident, 34
Adanson, M., 5, 8, 10, 43, 47, 51, 56
Africa, 136
Age-classes, 123–124, 128
Age-group systems, 136, 139, 151
Alternate linking, 136, 139, 151
Alternation, 14–16, 62, 121–154;
    definition of, 124–125; equation
    for, 127; formal account of,
    144–147
Analogy, 93, 94, 121
Année sociologique, 65, 88
Aphorisms, 3, 4, 34, 35, 38 n. 1, 62
Apo Kayan, 76
Aranda, 130–132, 137, 138, 147, 149
Araucanians, 54
Archetype, 47, 152
Argentina, 98
Argumentation, 3, 4, 19, 21–22, 23,
    32, 35
Aristotle, 83, 94
Aspect, change of. *See* Change of
    aspect
Asymmetry, 15, 93, 94, 122, 133.
    See also Prescriptive systems,
    asymmetric
Australia, 55, 130–133, 139, 140 n. 1,
    149, 150
Avestan, 156
Axiomatics, 14

Axioms, 2
Ayer, A. J., 110–111

Babcock, B., 105, 109, 110
Bacteriology, 48, 51, 53, 57, 59, 60
Balance, 47, 85
Bali akang, 73, 77 n. 6
Bambrough, R., 43 n. 7
Baram river, 69, 75, 76
Barden, G., 10
Barnes, R. H., 97, 103
Baron, A. G., 42 n. 5
Basic predicates, 6, 63, 64
Basle, 29
Batak, 98
Batek Negritos, 7
Bateson, G., 133, 135, 161
Beautiful, the, 8
Beckner, M., 44, 45, 46, 49, 50, 51, 58
Belief, ix, 6, 40–41, 65, 91
Bilateral marriage, 142
Biology, 51
Blanché, R., 2, 13, 14, 145
Blessing, 89 n. 14
Blood, 72
Borderline cases, 45, 58
Borges, J. L., 3, 34, 36
Borneo, 12, 68, 69, 72, 99, 128. *See
    also* Sarawak
Bororo, 99, 103

Botany, 43, 51, 60, 127
Boundaries, 118
Bow, 155–156
Bowmen, Persian, 123
Brazil, 98
Broad-arrow, 163, 164
Browne, Sir Thomas, ix, 38 n. 1, 98,
    105, 121, 164
Buginese, 128
Bühler, K., 38 n. 2
Bungan cult, 76

California, University of, 42
Campbell, K., 38 n. 1, 63
Cannibals, 95, 100
Castaneda, C., 31
Celebes, 69
Centers of variation, 11
Cerebration, 118
Certainty, 29–30, 31
Chain complex, 37, 47, 112
Chance, 33
Change of aspect, 2–3, 16, 17,
    22–23, 34, 120, 148, 160, 164
Change of sign, 119
Characteristic features, 98, 101,
    102, 112
Chawte, 165, 166
Chinese, 126, 156
Chukchee, 99
Class, conceptual, 6
Classification, monothetic. See
    Monothetic classification
Classification, polythetic. See
    Polythetic classification
Cluster analysis, 10, 56, 57
Cog-wheels, 90
Cogito, 33
Collective representations,
    definition of, 31–32
Collingwood, R. G., 30
Colony morphology, 59
Complementarity, 15, 62
Complementary governance, 122
Complementary predicates, 110–
    111
Complex thinking, 37
Computers, 47, 51, 56, 57, 59

Concretization, 104
Continuity, 104
Convergence, 149, 150
Cousin terminology, 53, 54, 55
Crevel, R., 9 n. 2
Cursing, 85 n. 14
Cycle, 127

D'Alembert, J. Le Rond, 9
Darius, 123
Dayak, 68. See also Ngaju, Penan
Death, 164
Death-names, 128–129
Descartes, R., 33
Descent, 52–53
Descent systems, 39–40, 51, 55, 64
Devil, 98, 103
Diagrams, 147–148, 155, 157, 164,
    165, 166
Dichotomy, 144, 153. See also Duality
Dieri, 132–133, 137, 138, 148, 149
Difference, 59–60
Diffusion, 53 n. 9, 139
Disorder, 103, 106
Downs, R. E., 70
Duality, 4, 139, 144, 145, 146, 151
Dublin zoo, 11
Duck-rabbit, 120
Dumont, L., 140 n. 1
Durkheim, E., 20, 25, 87, 103

Ecuador, 80 n. 9
Electricity, 86, 87
Elementary structures of kinship,
    165
Elshout, J. M., 72–75, 77, 79, 85
Endicott, K. M., 7
Energy, 87, 88
Enlightenment, the, 65
Epistemology, 35, 83, 122
Eros, 156
Escape reactions, 45
Escher, M. C., 119
Evans-Pritchard, E. E., 62
Evolution, 43, 51, 52–56
Exemplary scenes, 4
Extremity, 108
Ezekiel, 155

Fact, concept of, 20−21
Family resemblances, 5, 9 n. 2, 37 n. 1, 38 n. 3, 41, 45, 63, 65. *See also* Serial likenesses; Sporadic resemblances
Fibers, 47
Flood, Mr., 11
Flores, 22
Flute music, 136, 149, 150
Formal analysis, 62, 63, 113−117, 160
Formulas of reversal, 115−116
Fortune, R. F., 142
Four-section system, 129. *See also* Kariera
Frazer, J. G., 139
Frege, G., 110
Frieze, 123, 124, 150
Fürer-Haimendorf, C. von, 70
Furness, W. H., 70

Garo, 55
Gauss, C. F., 35
Genealogical specifications, 141
Generality, craving for, 91
Generalized exchange, 140, 166
Gennep, A. van, 87
Gift, 87
Gilmour, J. S. L., 43 n. 6
Global distribution, 152, 153
God, 33, 36
Golden Rose, the, 29
Gradient, 29, 30
Grimm, J., 38 n. 1

Hair, 69
Half-man, 152
Hallpike, C. R., 106
Harmonic regimes, 140
Harner, M. J., 79 n. 9, 90 n. 15
Hartman, P. A., 48, 53, 57, 59
Hawä, 101
Hawaii, 14
Head-hunting, 11−12, 66−92
Heraldry, 158, 163
Hierarchy, taxonomic, 46, 48, 50
Homology, 93, 94, 121
Hose, C., 69
Hostility, 136, 139

Huichol, 98, 100, 102, 103−104
Human nature, 40, 91, 106
Humbert, H., 119
Hume, D., 83
Hussey, E. L., xiv
Hutton, J. H., 71, 84, 85, 86
Huxley, F., xiv

Iatmul (Iatmül), 54, 133−136, 137, 138, 143, 147, 149, 150, 152
Iban, 76, 78 n. 7
Incest, 52, 95, 96, 100, 108
India, 55, 98, 157
Indonesia, 22, 55, 71 n. 2, 100, 128−129
Initiation, 135
Inner states, 6
Interpretation, 23, 28, 29, 30, 31, 32, 74, 75, 83, 120, 161, 165
Intuition, 13
Inversion, definition of, 95−96. *See also* Reversal
Ireland, 98
Iron, 72
Izikowitz, K. G., 66, 80, 85, 86

Japan, 98, 107
Japanese, 76
Jevons, W. S., 37, 38, 41
Jívaro, 12, 79 n. 9, 90 n. 15

Kaguru, 98, 109
Kāma, 156
Kant, I., 66, 83, 144
Kariera, 129, 147
Karlgren, B., 126
Kédang, 96−97, 98, 103
Kenya, 123
Kenyah, 69, 72−79, 81, 82, 83, 99
Key classification, 46, 50
Klpu, 66−68, 79, 82 n. 10, 85, 86
Knight, R. Payne, 9
Konyak, 70
Korn, F., xiii, 133, 135, 137−139, 140, 148, 149
Kruyt, A. C., 68−70, 71, 85, 89, 109
Kükenthal, W., 69

La Saussaye, P. D. Chantepie de, 69
Ladder, 5
Lamet, 66–68, 79, 80, 82 n. 10, 85
Laos, 66
Lévi-Strauss, C., 137, 138, 139–142, 165–166
Levitation, 119
Lévy-Bruhl, L., 82, 83
Lichtenberg, G. C., 5, 8, 10, 33–35, 38 n. 1
Life-energy, 71, 78, 85
Life-fertilizer, 71, 81
Life-force, 70, 72, 81, 85
Life-fluid, 69, 85
Lightning, 34
Linnaeus, C., 43, 46
Lions, 11
Lockhart, W. R., 48, 53, 57, 59
Long Muh, 76, 78
Long Nawang, 77
Long San, 76
Lugbara, 95, 97, 100, 107, 109
Lukes, S., 87 n. 12

MacDonald, M., 75 n. 5
Magic, 72, 74, 75, 85
Magritte, R., 119, 160
Makassarese, 128
Malaya, 98
Malaysia, 7
Mana, 80 n. 9
Manggarai, 22, 55
Mapék Arang, 77
Mara, 137, 138
Marriage, 58–59
Marriage cycles, 165–166
Matrilateral cross-cousin, 140–141
Mauss, M., 87
Mawas, 98
Maxims, 3, 35
Mayr, E., 50, 51 n. 8, 58
Mbapma, 133, 135, 149
Mechanics, 86, 87
Mental reality, 139–140, 143
Meru, 123–124, 128
Metaphor, 3, 31, 47, 92, 96, 104, 106, 107, 108, 112–113
Middleton, J. F. M., 95

Mirror, 157, 159, 160
Miwok, 54
Module, alternating, 137, 149
Monothetic classification, 10, 13, 47, 64, 65, 112, 117; definition of, 48–49
Monotypic, 44–45, 47
Morris, M., 3
Mpuga Rukidi, 100, 110
Mugwe, 124, 128
Muisak, 78–80 n. 9
Murdock, G. P., 53–55
Murngin, 140
Myerhoff, B., 103–105

Naga, 68, 79, 81. See also Konyak
Natural kinds, 18
Natural resemblances, 91
Natural selection, 43
Naven, 99
Necklace, 14
Needham, T. R. A., v, xi, xiv
Negation, 110–111, 114; cultural, 105, 106, 110
New Caledonia, 98
New Guinea, 133
Ngaju, 99, 101
Nietzsche, F., 37 n. 1
North America, 137
Numerical taxonomy, 5, 11, 47, 56, 60. See also Cluster analysis
Nyoro, 54, 97, 100

Occam's Razor, 81, 84
Odd-job word, 58
Ogden, C. K., 94
"Omaha," 55, 62
Open texture, 121
Opposition, 14, 16, 17, 93, 94, 95, 97, 102, 104, 108, 111–112, 121, 122, 143; definition of, 96; Lévi-Strauss on, 139
Order, 30
Ordnance, Board of, 164

Paradigm, 23, 24, 30, 31, 32, 35, 62, 113, 158–159
Paradox, 34, 119

Pau, R. N., 42 n. 5
Pārvatī, 157
Peacock, J. L., 105
Penan, 76, 77, 83 n. 11, 128–129
Penis, 157
Periodicity, dyadic/triadic, 150–151
Peyote, 100, 104
Phallus, 157
Pheon, 163
Philology, 17–18
Phylogeny, 53, 54, 57
Physics, 86, 87, 88, 92, 126–127
Picasso, P. R., 33, 34, 35
Plato, 83
Poincaré, J. H., 35
Polythetic classification, 5, 6–11,
    36–65; definition of, 49
Polytypic, 44–45, 46–47, 58
Portrait, 33
Prescriptive systems, x, 54, 55, 56,
    64, 93, 94, 122, 129–136, 148, 155;
    alternating, 137, 149; asymmetric,
    133, 138, 149, 161, 165–166;
    patrilateral, 142
Prestations, 154, 161–162
Primary factors of experience, 65, 153
Principle, 144, 145, 146, 148, 153; of
    organization, 95
Proclivities, mental, 64, 118, 119,
    122, 123, 126, 150, 151, 152,
    153, 154
Puerto Rico, 3
Puli-Akha, 68
Purifying, 74, 89 n. 14

Quiver, 162

Radcliffe-Brown, A. R., 139
Reasoning, 33, 40
Relationship terminologies,
    51–52, 64
Relationship terminology,
    Iatmül, 133
Representation, 16, 18, 31–32, 35
Reversal, 13–14; definition of, 96
Rhees, R., 11
Right/left, 99, 101, 107, 114, 117
Rigidity, 90, 91

Rites of passage, 87
Ritual, 89 n. 14
Riverside, 42 n. 5
Rivière, P. G., 58, 79 n. 9
Romans, 27
Rope, 37, 47
Rotation, 107, 160
Roti, 98, 100–101, 102, 109

St. Peter, 102
Sanskrit, 126, 156
Sarawak, 70, 75
Scientific idiom, 86–88
Sebop, 69 n. 1, 70
Secret sympathies, 18
Seeing, 22
Separation, categorical, 103, 104
Sepik river, 133
Sequence, 145–146
Serial likenesses, 38 n. 3, 40,
    51 n. 8, 52
Sextus Empiricus, ix
Similarity, 157–158, 165
Simplicity, criterion of, 153
Simpson, G. G., 46–47, 50, 58
Śiva, 157
Skull-rack, 77, 84
Smith, J. Z., 102
Snake, 98
Sneath, P. H. A., 44, 47, 48–49, 52,
    57, 58
Sokal, R. R., 44, 48–49, 52, 57
South America, 55, 137
Soul-substance, 68–72, 75, 78, 81,
    84, 91
Sporadic resemblances, 38, 39, 41,
    65, 112
Steinberg, S., 159
Stewart, D., 8–9, 10
Stewart, F. H., 136, 137, 139, 151
Streams, 139, 151
Structural analysis, 121, 154
Structure, 148, 165
Substitution, 37, 38, 56, 62
Sumatra, 98
Sumbawa, 128
Surveying, 163
Susa, 123

Syllogism, 152
Symbolic classification, 87, 105, 122
Symbolism, 28–29
Symmetry, 15, 62, 64, 93, 122, 165

Tanoana, 70
Tarau, 165, 166
Teknonyms, 128–129
Thado, 54
Thing, 20–21
Tiffauges, A., 119
Time, linear, 104
Tinjar river, 70
Tokong, 69
Tone of thought, 86, 89, 92
Toradja (Toraja), 69, 70, 79, 81, 98, 99,
    100, 101–102, 107, 108, 109, 117
Toulmin, S., 38 n. 2
Tournier, M., 119
Transformation, 12, 104, 105,
    113, 115
Transitive expression, 9
Transitivity, 4, 15, 62, 64, 93, 122,
    153; definition of, 94
Translation, 31, 123
Transposition, 107, 108, 116, 117
Transvestism, 99, 107–108, 110, 114
Triangle, 23
Tsantsa, 80 n. 9
Two-line terminologies, 55, 129

Two-line system, transformation to
    four-line, 148–149

Uganda, 95
Unconscious, the, 122
Unison, 146

Vagina dentata, 152
Value, 21
Vulva, 157
Vygotsky, L. S., 5, 6, 8, 37, 47, 90, 112

Wa, 68
Wagner, R., 1
Waismann, F., ix, 33, 34, 35, 86, 91
Weyl, H., 123
Wisdom, J., 11
Witch, synthetic image of, 153
Witches, 98, 109
Wittgenstein, L., ix, x, 5, 6, 37, 41, 42,
    45–46, 47, 51, 58, 84, 90, 93, 119–
    120; on games, 11; on description,
    22–24; and Lichtenberg, 38 n. 1;
    on defective definition of class, 91;
    on negation, 110; on similarity,
    157–158
Words, 17–18, 65. *See also* Philology

Zoology, 42, 50, 51, 52, 55, 56, 60